THE FUTURE IS FUNGI

Dedicated to Dot

ILLUSTRATION NOTE

Scientists estimate that over 6 million fungal species exist in nature. To date, only 120,000 species of fungi have been identified, which means that 98 per cent are still to be discovered. We took creative licence with some of the illustrations to imagine the diversity of forms that could exist in the fungi kingdom.

The cover image is an imagined mushroom, reflecting fungi's role as creative architects and alchemists in the natural world.

The chapter opener images portray organic landscapes that propose possible futures, or alternative presents, in which fungi have continued to colonise and thrive. They depict a world where fungi's indispensable role in our universe is seen and understood.

The mushroom profile images illustrated on black backgrounds have been modelled as scientifically accurate representations of known fungal species.

All illustrations in this book, except two diagrams, were created by Joana Huguenin as 3D artworks, using mathematical models to simulate various forms before applying textures, colours and lighting. The digital approach reflects the marriage of nature and technology that's helping us discover, catalogue and understand all of the fungal species in nature.

MICHAEL LIM   YUN SHU

ILLUSTRATED BY JOANA HUGUENIN

THE FUTURE IS FUNGI

T&H

FUNGI SHAPE OUR WORLD

1

THE KINGDOM OF FUNGI 16

2

FOOD 42

3

4

5

MEDICINE 80

PSYCHEDELICS 116

ENVIRONMENT 164

Foreword

DR GUNTHER M WEIL

One of the core members
of the Harvard Psilocybin
Project in 1960-1963,
Dr Gunther Weil worked
closely with Timothy Leary,
Richard Alpert (aka Ram
Dass), Ralph Metzner and
George Litwin on pioneering
psychedelics research. He
received his doctorate from
Harvard University in 1965.
He was a Fulbright Scholar
in Europe and was recruited
by Abraham Maslow to
teach at Brandeis University.
He is the founder and CEO
of Value Mentors and an
internationally recognised
organisational consultant,
executive coach and
educator that advises leaders
and their organisations on
values-based leadership,
innovation, team building,
and executive wellness.

I was twenty-three when I had my first
psilocybin experience in the autumn
of 1960. The setting was the plush
living room of a mansion in Newton,
Massachusetts, leased by my Harvard
University faculty adviser, Dr Timothy
Leary. Tim was forty years old, in his
second year of a faculty appointment
as a lecturer in clinical psychology
in what was then the Harvard
Department of Social Relations. In the
first year of our Harvard Psilocybin
Research Project, this living room,
decorated with Indian prints, images
of the sacred arts and an audiophile-
quality sound system, was the hub for
near-weekly psilocybin sessions that
drew many interesting characters.

On the day of my virgin
psychedelic experience, I shared
the company of Matthew Huxley
(Aldous Huxley's son), Allen Ginsberg
and his partner Peter Orlovsky,
Dr Richard Alpert (later known
as Ram Dass), Ralph Metzner, Tim,
and his close friend and psychologist
colleague, Frank Barron. My first
psilocybin experience was deeply
transformational and altered the
trajectory of my life.

The psilocybin session was a
deep experience of unitive (non-dual)
consciousness: that is, the quantum
interconnectedness of all phenomena.
I explored multiple layers and levels
of this consciousness. I realised that
my life experiences up to that point
were just a very small fraction of the
range of possibilities offered by deeper
lenses of consciousness. I was able
to disconnect from the ego-centred,
little-me self and see more objectively,
and with bemusement, the 'Gunther'

character and how he operated through
his ethnic and culturally inherited,
conditioned world view – which
I define as the unconscious lenses,
beliefs or assumptions through which
we look at life.

These experiences were a massive
turning point in my world view.
I observed the unconscious framework
of how my life was run, programmed
by a variety of life experiences,
including childhood traumas.
Everything I had learned and believed
up to that point got turned inside out.

In our first meeting, Tim,
nattily dressed in a professorial
tweed sport coat, khaki pants and
sneakers, enthusiastically shared his
transformational epiphanies and
revelations from his first psilocybin
experience the previous summer in
Cuernavaca, Mexico. Since then, he
had devoted himself to studying the
role that psychedelics could play in
understanding consciousness and
the human brain, nervous system
and psychology.

Tim was boyishly handsome,
seductively charming, witty and very
smart. He graciously said, 'If you're
interested, then I'm happy to be your
advisor. If you're not, you're better
off working with someone else.'

I had an immediate, intuitive
response. 'Count me in!' With that
declaration I became one of the core
Harvard Psilocybin Project team
members. The team included Ralph
Metzner and George Litwin, who
were also graduate students, and
Dr Richard Alpert. To prepare myself
for this new adventure, I started
reading about the history of sacred

fungi and plants, their use in ancient rituals, such as the Greek Eleusinian Mysteries and Indigenous shamanistic cultures from around the world, and early mescalin-based psychedelic explorations by European artists.

I later realised how my upbringing, education and life experiences primed me to meet Tim and join the Harvard Psilocybin Project team. My parents and I fled Nazi Germany in late 1939 on one of the last immigration boats. I was two years old. In the United States, I was raised in a middle-class, academic and musically gifted Jewish family that supported my intellectual and creative interests. I was always highly curious and inquisitive. In fact, to this day, one of my core values centres on searching for life's meaning. I grew up in the bebop era in the 1950s, hanging out in jazz clubs in Milwaukee and Chicago in my mid and late teens. I was introduced to cannabis by my jazz musician friends and experienced many enhanced cognitive and sensory flow states and epiphanies. I graduated from Kenyon College in 1959 with a dual major in philosophy and psychology. In the academic interim year of 1959 to 1960, before entering graduate school at Harvard, I was awarded a Fulbright scholarship at the University of Oslo in Norway. Throughout that year I spent extended periods of time in Paris, hanging out in the jazz scene with a variety of expat American artists.

During my first three years at Harvard, I was involved in many psilocybin sessions. It was early on and we were just trying to understand the multiple dimensions of the psychedelic experience. It was like visiting another planet, observing and participating at the same time, and then writing about it like anthropologists. We were essentially learning 'on the job'. Our individual and collective shared purpose was discovering what these substances could offer with respect to the expansion of consciousness for healing and for scientific and artistic discovery.

Based on our study of ancient and Indigenous rituals, we quickly realised that 'set and setting' were critical variables to support a safe and illuminating psychedelic experience. We defined 'set' as the internal intentions, beliefs, expectations and mindset that people bring to the psychedelic experience. We defined 'setting' as the physical space, atmosphere, location and aesthetics of the environment; it's the external sensory input – the sounds, smells, visuals, etc.

In order to maximise a positive transformational experience, we needed to minimise fear and anxiety in the participants. This involved creating a comfortable setting of sensory and cognitive ease and emotional safety. It was clear to us that everyone had their own healing intelligence inside them, so we understood that, generally, 'less is more' when guiding psychedelic sessions. We were learning to guide by intuition, informed directly by our own personal experiences.

We were explorers who were on the cutting edge of psychedelic research. We embraced the belief that what we were doing was extremely important for understanding human consciousness, and that we were destined to change and raise consciousness for the betterment of all. We were convinced that we had the ultimate tool, declaring psilocybin to be a high-powered electron microscope that could be used to enter into hidden dimensions of consciousness.

Tim, by nature a rebellious Black Irish poet, was intellectually daring, contemptuous of dogma and arbitrary institutional authority, and had an unflinching faith that psychedelics were going to change the world. He was acutely aware that great prophets are usually not understood or welcomed in their time.

When Tim and Richard Alpert were fired by Harvard in 1963, they moved to Millbrook, a historic sixty-four-room estate in upstate New York on 2300 acres, rented to Tim by Peggy, Billy and Tommy Hitchcock, heirs to the Mellon family fortune. Tim and Richard attempted to continue their psychedelic research there under the rubric of the League for Spiritual Discovery and the Castalia Foundation.

Since psychedelics had become taboo at Harvard, I had to switch my thesis focus in order to graduate the following year. Upon graduation in 1965, I accepted a teaching role at Brandeis University offered to me by Abraham Maslow, known by most for his 'hierarchy of needs' and widely considered the father of humanistic psychology. Abe was studying peak experiences, and since I was in the business of creating them, we bonded. A year into my multiple year contract, much to Abe's disappointment, I felt the urge to join my old Harvard team. In June 1966, I resigned my Brandeis faculty appointment and moved my wife and two children to Millbrook to join Tim and Richard.

It was an exciting and seminal time at Millbrook. Streams of people from the arts, sciences, literature and many other walks of life came through the large stone gateway towers of the estate. They included personages such as Alan Watts, Allen Ginsberg, Dizzy Gillespie and Charles Mingus. They tended to be people who were creative and adventurous risk-takers. That's not to say the Millbrook experience was easy. Nina Graboi, an early close supporter and collaborator in the League for Spiritual Discovery, described Millbrook as 'a cross between a country club, a madhouse, a research institute, a monastery, and a Fellini movie set'.[1] My wife and I agreed that Millbrook wasn't an environment to raise our children in.

One night, a month or so after arriving in Millbrook, I had a highly lucid, prophetic dream. I immediately knew that if I stayed there it would be destructive to myself and my family. I was a 27-year-old husband and a father of two young children, and I felt the need to resume my professional life. After only several months, I left Millbrook to return to Cambridge and resume an ordinary life. Subsequently, my personal and professional involvement with psychedelics faded away and was replaced over time by a variety of psychospiritual and energetic disciplines, such as meditation, qi gong and tai chi.

I believe I am the last representative of the original core Harvard Psilocybin Project team. Given this legacy, I hope to share my learnings and insights as I reflect back on the psychedelic experiences of my early life. Who would have thought that these explorations were made possible by the kingdom of fungi: that the precursor to LSD lay hidden within the ergot fungus, and that the inconspicuous psilocybin mushroom grows ubiquitously in the wild around the world.

One of my major life learnings is the understanding that whenever a massive shift in world views occurs, the associated values of the new world view accompany it. In my life, I embrace values such as love, transcendence, oneness, compassion, empathy and collaboration. But it is very important to understand that while psychedelics clearly have the potential to temporarily open a portal – they can show you the possibilities of abiding in inner freedom – they do not sustainably establish that freedom. Essentially, you have a few hours of experiencing deeper insights and a similitude of inner freedom. I was able to see the beautiful mystery and the pervasive love that supports the manifested universe, including this gigantic movie that we're all in. However, it is dangerous to rely on psychedelics to keep repeating a certain kind of spiritual or transcendent experience. You can easily get trapped pursuing endless psychedelic experiences, which becomes a dangerous spiritual and psychological dead end. The real work of integrating psychospiritual insights and a more generative world view and its associated values and behaviour begins after the psychedelic session.

This integration involves deconstructing faulty beliefs, attitudes and assumptions, and healing the associated traumas in the body and mind. Spiritual work helps, but it cannot handle the entire task and process alone.

For me, there is an unmistakable direct line from my Harvard psychedelic experiences of my mid-twenties to my sixty years of inner work with psychospiritual disciplines including the Gurdjieff Work, Buddhist, Taoist and Vedic Advaita teachings and practices, as well as the martial and healing arts of qi gong and tai chi. It's only in the last few years, stimulated by the current psychedelic renaissance, that I have explored the psilocybin journey again, but with very specific intentions. At this stage of my life, approaching eighty-five, I'm exploring my felt sense of mortality while continuing to heal some residual traumatic elements stemming from the Holocaust experiences of my parents and myself when I was a child.

I have devoted many years of my life to exploring human consciousness. But psychedelics are just one of the phenomenal arcs of fungal creation. They also shape the natural world, produce life-saving medicine, and create some of our favourite cuisines. Interestingly, if I map the fungi kingdom to Maslow's hierarchy of needs, fungi can help us meet many of our needs. Fungi produce food, medicine and raw materials for our basic physiological requirements. They don't just meet and exceed our safety needs, they also offer innumerable biotechnological and mycoremediation solutions for decontaminating the environment. Psychedelics take us on a journey of inner worlds to experience conscious love and self-transcendence of the 'little me'. Fungi have wideranging potential that permeates the human experience, and this book has placed the human–fungus connection at the centre of all of it.

Michael and Yun have created a fascinating, refreshing and infectiously thought-provoking exploration of fungi. They find keys to the future by following the evolutionary roots of fungi from the past to modern-day research and applications. There is wisdom to be found in these pages as they share their wonder and awe about a billion-year-old kingdom that has refined its survival strategies in a vast, collaborative network of deep ecology.

This book is a rediscovery of our relationship with ourselves, nature and the world around us. Only then can we realistically and practically explore the possibility of co-creating a future that is as enduring, resilient and creative as that of the fungi kingdom.

Psychedelics take us on a journey of inner worlds to experience conscious love and self-transcendence of the 'little me'.

Introduction

MICHAEL LIM
YUN SHU

Nature takes her time. Without rush, sublime and elegant life forms flourish into their truest expression. Nature is creativity unfolding – sculpting, trading, transforming, decomposing, recycling. The kingdom of fungi embodies the pinnacle of this design, underpinning life on Earth for over a billion years. With each fine thread of mycelium, fungi weave the tapestry of the living world.

Fungi have long been the architects and alchemists of the natural world. For thousands of years, humans have harnessed their molecular powers to turn wheat into bread, and fruit into wine – and our lives are richer for it. Advances in technology since the mid-20th century have accelerated the study of fungi – mycology – into a biotechnology revolution that has endless potential to enrich our world.

In these pages, we explore how fungi can be leveraged to transform matter into a diverse set of radical solutions for today's urgent ecological and social issues. Fungi are being engineered to increase food production, grow future meat alternatives, create new sources of medicine, produce sustainable biomaterials, remediate the environment and change our state of consciousness.

Fungi traverse every corner of the globe and can even survive in outer space. They provide endless lessons in ecology that defy our traditional notions of what it means to be intelligent. Fungi redefine resourcefulness, collaboration, resilience and symbiosis. Fungi have survived all five mass extinction events and provide innovative solutions for thriving on our planet. Yet we know much less about fungi than we do about the animal and plant kingdoms.

The future of fungi permeates all disciplines and business models. Their allure not only lies with researchers uncovering applications for the future but extends to capitalists creating value in the burgeoning fields of food and medicine biotechnology, mental health services and environmental remediation. A world of discoveries awaits as the wild and mysterious world of fungi is revealed to us.

Fungi, and mushrooms in particular, are entering the cultural zeitgeist. Foraging for edible mushrooms, cultivating medicinal mushrooms at home, using psychedelics for personal liberation, and reconnecting with shamanic rituals are practices that are emerging from the underground. The future is now, and it's being built by people who are learning about the changes that need to happen – changes that can be catalysed by fungi.

The Future is Fungi transpired from our enduring fascination with fungi and the human condition. In true mycelial style, discussions about fungi quickly radiated into an exploration of history, science, philosophy, spirituality, holistic wellbeing and ecology. We are driven by a desire to speak for a hidden yet extraordinary kingdom that doesn't express itself using words. This is a book about fungi and a future that embraces their wisdom. It is also an invitation to discover a deeper awareness of yourself. To be a breathing, conscious being means that it is our obligation to make sense of the world and our role in it. We hope to share our anthropological curiosity by using fungi as a vehicle to investigate and reckon with our role in this small slice of time and space.

Perfection is the enemy of progress, so this is by no means an exhaustive exploration. This is simply one of many expeditions using fungi to explore the far reaches of our world and our minds. Nature gave us the magic of consciousness, so we hope that you find a new concept in this book, give it a spin and observe it.

Fungi teach us that we are all interdependent. When we finally surrender our separateness, we realise that we are not outside of nature, but within it. Welcome home.

The largest living organism in the world is a fungus. *Armillaria ostoyae*, commonly known as the honey fungus, grows in the form of mycelium underneath Malheur National Forest in Oregon.[1] The honey fungus covers an astounding 965 hectares – equivalent to 1600 soccer fields – and, based on its current growth rate, has been successfully expanding for up to 8000 years.

FUNGI SHAPE OUR WORLD

Nothing alive exists in isolation. To be alive means to be part of an intricate, sprawling web of cause and effect. Our lives are interwoven with that of plants, animals, bacteria and fungi, forming the beating heart of the planet we are in partnership with.

The most underappreciated and under-researched of these threads of life is the kingdom of fungi. Fungi are incredibly powerful and complex and their omnipresent role in nature is still widely unrecognised. As light shines on this mysterious world, we believe that future research and investment into this space will only grow. Rapidly.

Fungi shape and transform environments, under-pinning the wellbeing of nearly all terrestrial ecosystems. Despite spending most of their lives hidden underground or inside plants and animals, fungi are responsible for critical ecosystem processes. Some weave through the earth, decomposing matter and recycling nutrients to build healthy soils for plants and animals to flourish in. They are the interface between death and life – without them, the world would be buried under fallen trees, the remains of animals and infertile soil. Others form intimate and intelligent partnerships with all forms of life, supporting the health of most – if not all – organisms. Modern-day privileges such as beer, wine, chocolate, bread, penicillin and detergent depend on fungi for their production, and a potent group also contains psychoactive compounds that can initiate transformative experiences of love, creativity and connection.

It is safe to say that without fungi, the world as we know it would not exist. Fungi are nature's alchemists and hold untold answers for our future on Earth.

1.1

WHERE DID WE ALL COME FROM?

To best appreciate the importance of fungi, let's start at the intersection of science and history.

Our Earth is 4.5 billion years old. The tree of life can be traced back to LUCA – the Last Universal Common Ancestor – some 4 billion years ago.[2] The earliest living organisms were basic, single-celled prokaryotes. Single-celled organisms are the smallest thing you can be and still be considered a living organism. They are represented by the domains *Archaea* and *Bacteria*. Trees, birds, insects and fish, as we know them today, did not exist.

Over millions of years, to survive the changes in Earth's conditions, prokaryotes fused into larger and more complex living organisms called eukaryotes, enabling the diversity of nature that we see today. Eukaryotes form the domain *Eukaryota*, which is organised into four kingdoms: *Plantae* (plants), *Animalia* (animals), *Fungi* and *Protista* (everything else). Humans are mammals, and sit within the kingdom of *Animalia*, along with all other animals from electric eels to elephants. It is humbling to contemplate that all life, be it an oak tree or Gandhi, descended from the same primitive organisms.

Fungi diverged from animals 1 billion years ago. So close are the *Fungi* and *Animalia* kingdoms that taxonomists speak of a super-kingdom that combines the two: *Opisthokonta*. This billion-year-old divergence was only recognised in 1969 when ecologist Robert Harding Whittaker formalised their importance, scale and diversity with the new kingdom classification. Previously, fungi were classified as plants, and dismissed as lower-class organisms, tucked away in an obscure corner of the botany department.

As Whittaker recognised, fungi are more closely related to animals than plants. Say hello to your evolutionary cousin – we share almost 50 per cent of our DNA with fungi. The basic building blocks of fungal cell walls are chitin, the same material as the hard shells of crustaceans and insects. The cell walls of plants, in contrast, are built from cellulose. Like animals, and unlike plants, fungi are heterotrophs – they are unable to produce their own food and must instead obtain food from their environment. Animals opted to internalise their stomachs, while fungi pursued external ones. They secrete enzymes into the environment to digest food externally before absorbing it into their cells. Fungi also have a remarkably wider taste palette than

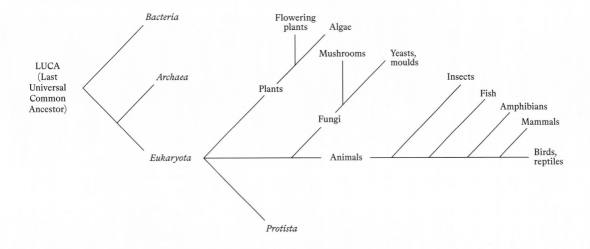

we do, consuming everything from stale bread to plastics and even nuclear waste. This ability is the basis for modern food and medicine biotechnology, and has applications in environmental remediation.

Fungi are inherently difficult to study in nature, as the majority of species are microscopic or underground, but mycologists have made great strides in recent decades. Technological advances make it possible to identify fungal DNA in the environment. Unfortunately, the DNA doesn't match any records because a comprehensive catalogue of fungal DNA does not yet exist. Even once catalogued, the species could sit in a fungarium for decades before funding is granted to understand their role in nature, their interactions and how they fit into the tree of life.

To date, only 120,000 species of fungi have been identified. Using a complex process called DNA barcoding, scientists estimate that over 6 million fungal species exist in nature, which means that 98 per cent are still to be discovered, highlighting the untapped potential of mycology.[3] Every year, a few thousand new species are described. At this rate, it will take more than a thousand years to identify all the fungi on Earth. Mycologists obviously aren't running out of new species to discover – the limiting factor is simply that there aren't enough mycologists. The practice of cultivating mushrooms, or mycoculture, also has an engaged citizen scientist community that pushes the field forward. Websites such as iNaturalist and Mushroom Observer allow amateur mycologists and enthusiasts to log the mushrooms that they find to generate data on fungi diversity and conservation.

We know substantially less about fungi than we do about animals and plants, but discoveries are underway. In fact, fungi rarely leave the news cycle. Fungi are a key player in the biotechnology revolution, plant-based food movement, health supplements industry, grassroots remediation and psychedelic renaissance.

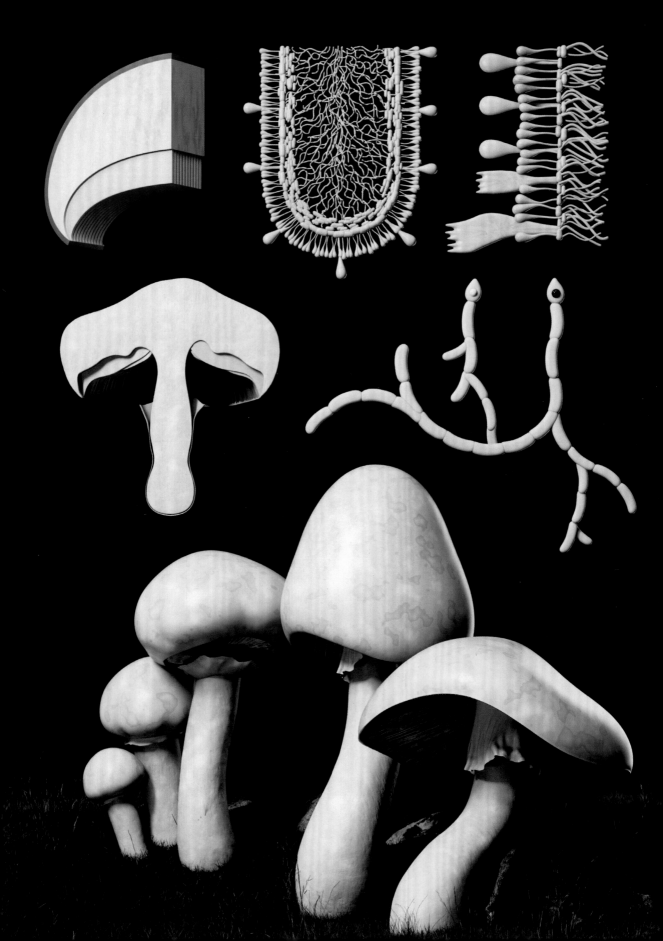

1.2

WHAT ARE FUNGI?

The kingdom of fungi is incredibly diverse. Beyond the popular images of cap and stem mushrooms, there are clubs, corals, shells and balls, just to name a few. And mushrooms are just the tip of the metaphorical iceberg.

In fact, only 10 per cent of (known) fungi produce mushrooms. Yeasts, moulds and mildews are all fungi, yet most of these are invisible to the naked eye and don't produce mushrooms at all. Mushrooms are simply the reproductive organ of an extensive fungal network. Thinking of fungi as being only mushrooms is like thinking of people as being only their genitals.

Turn over a pile of leaves in a forest and you'll almost certainly see white, furry patches hitched to their underside. These white, cottony masses are mycelia, which are made up of individual strands of thread-like hyphae. Mycelium and hyphae compose the growing and feeding part of the fungus – also known as the vegetative stage. This structure typically exists within soils. It allows the fungus to search for food but it also creates and aerates soils that benefit the rest of the ecosystem. Hyphae are most visible when they fuse, flood with water and produce mushrooms, which are recognisable as puffballs, morels, death caps and so on.

Fungi encompass both the mushroom and the mycelium. Repeat after me: all mushrooms are fungi, but not all fungi are mushrooms.

Mushrooms serve one biological purpose: reproduction. They contain spores, the reproductive units of fungi that function like the seeds of plants. Mushrooms come in various shapes, colours and sizes, but they all have just one function – to release their spores far and wide.

Mushrooms are referred to as sporophores, sporing bodies and, most recognisably, fruiting bodies. Because fungi were once studied as part of botany, many terms were borrowed from plants and these continue to be widely used. In this book, we will refer to mushrooms as sporing bodies to encourage the transition towards fungi-specific language.

Species of fungi that produce mushrooms for reproduction are macrofungi. The overwhelming majority that don't form sporing bodies are called microfungi. Aptly named, their reproduction process is typically invisible to the naked eye. To adapt to constantly changing environments, fungi other than yeasts maximise their survival by shuffling the genetic deck of cards. Through sex, these fungi develop new forms, features and chemistries.

Sex in the fungi kingdom is genuinely remarkable. If you peer at a fungus under a microscope, there is no sexual differentiation. There

LEFT

Clockwise from top left: Segment of a mushroom cap and gills; Magnified view of a gill; Highly magnified view of a gill releasing spores; Two suitable mating types of hyphae fusing to create mycelium; The growth cycle of mushroom sporing bodies; Cross-section of a mushroom sporing body.

are no males or females, no sex organs and no way to discern one fungus's sex from another. Mating is controlled by a number of genes that splits fungi into mating types.

Tom May, the principal mycologist at the Royal Botanic Gardens Victoria, Australia, describes it like this: 'You're at a nightclub and everyone's wearing a t-shirt with a number on it. Everyone looks the same in terms of gender because there's no sexual differentiation. As long as someone's number is different to yours, you can hook up with them. If the number is the same, then they're family and there's no possibility of successful mating.'

This mating approach gives fungi an immensely greater potential for genetic diversity. And this is the purpose of reproduction in the first place: to ensure your lineage becomes stronger, fitter and better adapted to the environment.

All mushrooms are fungi, but not all fungi are mushrooms.

1.3

HOW DO FUNGI FEED?

The classic food chain model distinguishes plants as producers, animals as consumers and fungi as recyclers. Mycelia shapeshift in a relentless search for food, water and nutrients for growth. They are nature's intermediaries, carrying out the processes between death and life.

Fungi respond to cues in the environment, such as sources of food, competition and pathogens, using sophisticated chemical signalling, although we know very little about how it works.[4] They can writhe, adapt and mould themselves in the most inhospitable environments to decompose, extract nutrients and bring about new life. To best appreciate how fungi have come to fulfil this ecological niche, let's go back in evolutionary history.

As land creatures, we rarely think about the billions of years of evolution that first occurred in oceans. Life in the ocean thrived in abundance, as early organisms danced in a hearty soup of nutrients and water. In contrast, the land was barren and scattered with rocky formations.

Scientists believe that microbial communities of fungi, bacteria and perhaps algae first lived on land 1 billion years ago.[5] Fungi survived by using penetrative and chemical strategies to mine minerals and water from rocks, which slowly disassembled the dense, complex structures into soil. They were the earliest colonists of a hostile landscape and they laid the groundwork for the ancestors of plants that arrived onshore a few hundred million years later.

These early plants were green algae, which had the ability to photosynthesise and produce their own sugars using the sun's energy. At first, they couldn't venture far from the ocean, as they had not yet formed root systems to access water and vital nutrients from the soil.[6]

A principle of nature is thrift – its processes happen in a way that requires the least effort. As competition for resources increased in the oceans, fungi and early plants formed a symbiotic partnership to ensure their mutual survival as they established themselves on the unchartered territory of land. An economic trade was forged: plants shared their sugars with fungi, while fungi shared minerals and water with the plants, acting as their root systems. This was a physical partnership – plants allowed mycelium to grow and occupy space inside their cells to facilitate the transfer. Fungi reciprocated by granting plants access to their mycelium. This was an easier way for plants to access micronutrients and water as they moved away from the ocean.

The lineage of fungi is not well documented in the fossil record, but it is widely accepted that this partnership occurred at least 480 million years ago.[7] Cross-sections of Rhynie chert – a sedimentary deposit with exceptionally preserved plant and fungus fossils from 410 million years ago – displays this intimate relationship, as mycelium can clearly be seen entering plant cells.[8]

Over millions of years, plants evolved their own root systems. They also formed lignin, the hard structural component in wood that allows trees to grow tall and dominate forest ecosystems. Fungi co-evolved with plants, and developed the unique ability to decompose lignin, capitalising on this new and abundant food source. This ability was instrumental to recycling matter – it allows the forest ecosystem to continue through cycles of birth, growth and decay. The atmosphere's composition also transformed as more plants inhaled carbon dioxide and pumped out oxygen, allowing for the growth of animals.

The long and enduring partnership between fungi and plants has fundamentally shaped the natural world and opened the door for the evolution of all plants and animals on land. Life doesn't always compete for resources as individuals. Symbiosis has also been a powerful tool for evolution. This is demonstrated by the benefits gained by both fungi and plants from intertwining their evolutionary histories.

Today, fungi have vibrant relationships with most members of the food chain and play vital roles in all ecosystems. Fungi and plants continue to partner, with more specialised relationships that reflect the complex modern landscape.

Fungi have evolved to feed on almost any organic matter using three broad strategies. Some form mutualistic relationships with plants or animals for food, benefitting both parties. If the relationship is long-term, it is also considered symbiotic. The largest groups of mutualists include mycorrhizal fungi, endophytic fungi and lichens. Another strategy is that employed by saprophytic fungi, which decompose dead or dying matter for energy. Finally, there are parasitic fungi, which feed off living hosts. Some fungi can employ more than one method or switch between methods as it suits them.

MUTUALISTS

Mycorrhizal fungi

Greek *mýkēs* 'of fungi' + *rhiza* 'root'

Mycorrhizal fungi are networks of underground mycelium that interact with the root systems of plants. They are the most widespread organisms in soil,[9] and as many as 92 per cent of all land plants have a mycorrhizal relationship with fungi.[10] Fungi share nutrients and water with plants in exchange for sugars. Although most modern plants are self-sufficient, their roots are much thicker than mycelium, making soil navigation to extract nutrients and water cumbersome. Mycelium is only one cell wall thick and can easily extract nutrients from a range of complex materials, breaking them into elementary substances such as water, carbon dioxide, nitrogen, phosphorus and calcium – all superfoods that allow plants to flourish. There are two connection methods: ectomycorrhizal fungi envelop the plant's roots, while endomycorrhizal fungi penetrate the cells of the roots.

Plants with mycorrhizal relationships grow stronger and can survive in previously inhospitable areas such as mountain ranges, the sub-Arctic tundra and tropical forests. In return for these benefits, fungi secure a reliable food source from plants.

Endophytic fungi

Greek *endon* 'within' + *phyton* 'plant'

Endophytic fungi are systems of mycelium that live and grow in plant tissue. All plants host one or more endophytic fungi in their cells that are critical to their microbiome – this is similar to the way that bacteria in the human gut make up our microbiome. In exchange for protection and food, endophytes produce chemicals that help plants absorb nutrients, resist diseases and tolerate stressful environmental conditions. These intimate systems of symbiosis are highly complex and are only starting to be understood.

Lichens

Greek *leikhēn* 'what eats around itself'

Lichens are a marriage between fungi and photosynthesising algae or bacteria. Fungi offer the algae or bacteria protection and shelter, and in return the fungi receive a steady source of nourishment. This union is the ultimate example of symbiosis – lichens live in habitats that neither fungi nor algae could survive alone. As such, lichens are studied as a single organism, pushing the boundaries of what constitutes an individual. New research suggests that a third critical player, yeasts, may also be entwined in this symbiosis.

It's not certain exactly when lichens evolved. Estimates range from 400 million years ago, during the early colonisation of land, to 250 million years ago, after plants had established themselves on land. As extremophiles – organisms that live in extreme conditions – lichens grow everywhere and dominate 8 per cent of the Earth's surface.[11] This area is more than the combined size of Canada and the US, or all of Europe. Visually unique yet unassuming, lichens challenge the notion of what it means to be alive. They are able to survive complete dehydration and lie dormant between life and death, in bleak conditions, only to come back to life once rehydrated. This enduring and sophisticated partnership is also of significant ecological importance: lichens break down rocks into soil, contribute to nutrient and water cycling, and provide nests and food for wildlife.

DECOMPOSERS

Saprophytic fungi

Greek *saprós* 'rotten, putrid' + *phyton* 'plant'

It is thanks to the thriving microbial world of fungi and bacteria that our world can regenerate at all. Saprophytic fungi are primary decomposers. They break down injured, dying or already dead matter into simpler compounds that can be used by the rest of the ecosystem. Where we see leaf litter, dead animals and fallen tree stumps, fungi see nutrient-dense food sources. Through this decomposition process, fungi recycle and make available nutrients that would otherwise be locked up in dead matter.

Saprophytes are the digestive tracts of the forest. Specifically, white rot, brown rot and soft rot saprophytes are responsible for breaking down wood, an exceptionally tough substance due to its structural components: cellulose, hemicellulose and lignin. The next time you see a log, take a closer look and notice the white threads of mycelium permeating the wood – this is the fungal recycling process in action.

PARASITES

Parasitic fungi

Greek *parásitos* '(person) eating at another's table'

Some fungi have pursued more aggressive strategies for survival. Parasitic fungi weaken or even kill living hosts by causing disease and infection. In humans, parasitic fungi usually enter through a wound or otherwise bypass a weakened immune system and cause conditions such as athlete's foot and ringworm. They receive plenty of negative media coverage due to their impacts on food crops, but many are beneficial for the food chain as they help balance populations in an ecosystem. Plants and animals that are killed by parasitic fungi provide nutrients that allow other organisms to live. There is no good or bad in nature.

Some genera of fungi, such as *Ophiocordyceps*, use eerie behavioural manipulation in insects to feed on and spread their spores. When spores of *Ophiocordyceps unilateralis* land on a carpenter ant, the fungus germinates within its body and secretes chemicals that hijack the ant's central nervous system and take control of its body. The fungus then instructs the ant to march up a tree to a branch where the temperature and humidity are optimal for the fungus to grow. The ant's final act is to bite down on a leaf to anchor itself while mycelium eats through its organs. Finally, a sporing body erupts through the back of the ant's head. Spores rain down and a new generation of evolutionary ingenuity continues.

RIGHT
Mycorrhizal fungi improve soil
health and plant growth.

WITH MYCORRHIZAL FUNGI

WITHOUT MYCORRHIZAL FUNGI

1.4

THE LIFE CYCLE
OF FUNGI

1. IT ALL STARTS WITH A SPORE

Spores are the beginning and end of the fungal life cycle. These single-celled units contain the code for a new fungal individual to flourish. Faced with countless microbial competitors and tough environmental conditions, the odds of germination are extremely low, so fungi release trillions of spores. They remain suspended in a state between life and death, monitoring the world around them for the right home. Spores are microscopic and so ubiquitous that there is no avoiding them – we all inhale ten spores with each breath.

A theory on the origin of life called panspermia even suggests that the blueprint for life was packaged in a spore and sent through space to find a home in the universe. While this is heavily debated, we do know that spores can tolerate extreme temperatures, resist radiation and even survive the vacuum of space. In 1988, Russian cosmonauts at the Mir space station noticed that something was growing on the outside of their titanium quartz window – and eating through it. It turned out to be a fungus.[12]

Like plants, most fungi are rooted into the soil and have slow reproduction methods, which involve physically growing their mycelium into new habitats or dispersing spores. Fuelled by the desire to propagate their DNA, some species of fungi have employed ingenious strategies to ensure their spores proliferate in new areas.

Black truffles (*Tuber melanosporum*), the fragrant culinary delight, is a prime example. This fungal gold grows below ground, and as its spores mature, it develops a fragrance that attracts animals, truffle hunters and gastronomists from around the world. Spores are not digestible. They eventually pass through the digestive tract of the lucky consumer, ideally a long distance away from the original location.

Above ground, the giant puffball (*Calvatia gigantea*) ripens millions of spores safely inside its round sporing body. To the delight of anyone who pokes it, it puffs out a smoke-like dusting of spores for the wind to carry away.

Pilobolus mushrooms grow in faeces and discharge their spores like a squirt gun by flooding their cells with water and building up pressure in the hyphae. Studies have calculated the spore projection at a rate of at least 20,000 g (g-force). By comparison, trained NASA astronauts wearing g-suits in a rocket are subjected to 3 g, and a bullet travels at an acceleration of 9000 g.

Other species of fungi bioluminesce, or glow in the dark, to attract insects to spread their spores across the forest floor. The coconut flower (*Neonothopanus gardneri*) is regulated by a circadian rhythm and glows brightly at night.[13] All this evolutionary tinkering ensures that reproduction continues.

2. FINDING A HOME FOR HYPHAE

When a spore lands in a place that is not too hot, not too cold, near food and near water, it will germinate. The spore absorbs water through the cell walls, and a thread-like tube called a hypha extends. As the hypha grows on a nutrient source, called the substrate, more hyphae branch out and create a string. The hyphae continue to eat through the substrate, which might be wood, insects or soil, and more hyphae grow from the tips, fusing and connecting to form an interconnected matrix called mycelium.

The growth of each hypha has combined physical force with a chemical strategy: hyphae secrete enzymes, which are strong digestive acids that break down matter. This makes fungi adept at penetrating through the toughest of substrates to extract the nutrients before transferring them into their mycelium. These acids are similar to those in our mouth – leave a piece of bread on your tongue and within seconds, enzymes in your saliva will turn the bread into a wet mush.

3. MILES OF MYCELIUM, AND MAYBE A MUSHROOM

Waves of mycelia radiate outward from the point where the spore germinated. The mycelium responds to chemical signals from nutrition sources and grows towards the food in a circular formation to maximise its surface area. When the food source in one area is exhausted, the inner mycelium is cannibalised and resources are redirected to the outer circle. Mycelium effectively grows as an expanding hollow ring. This is why you sometimes find 'fairy rings' in grasslands. The centre dies out as resources are redirected to the outer edges, gradually increasing the circumference of the ring. The mycelium can continue to grow in this way as long as nutrients and water are available.

During this stage, fungi other than yeasts can reproduce asexually by releasing spores from their mycelium. Microfungi such as moulds, rusts and mildew always reproduce this way – the black dots on bread mould contain upwards of 50,000 spores. Yeasts, however, take a different approach. They are single-celled microfungi that reproduce asexually by splitting off clones of themselves. Although they are efficient, yeasts miss out on the joys of genetic diversity that is ensured by sexual reproduction.

When environmental conditions are harsh – and they usually are – macrofungi can also reproduce sexually. When two sexually compatible mycelia meet, they fuse, forming a larger mass. This new mycelium, now genetically diverse, waits for the right conditions to compress its hyphae and inflate them with water to create a pinhead, known as a primordium. Over a few days, the primordium gradually elongates its stipe (the stalk-like part of the mushroom), which pushes the cap above the surface of the substrate. Finally, the cap opens, and it transforms into a fully grown mushroom. The mushroom can vary wildly in colour, texture and shape, depending on the species.

Macrofungi are typically split into two groups based on how the mushrooms produce and release spores. Ascomycota produce spores inside enclosed sacs. Basidiomycota form and release spores from gills, which are protected by a veil that peels away as the mushroom matures.

Mushrooms are a celebration – they signal a new generation of fungi, holding trillions of spores to be released. Spores will start the cycle again, as the process has gone on for a billion years. Nature is not sentimental; mushrooms decay the moment their work is done. They have served their purpose, but not without giving us a glimpse of the visceral beauty of nature. Maybe that's why mushrooms are so admired. They are the most beautiful moment in the fungal circle of life.

Mushrooms are a celebration – they signal a new generation of fungi, holding trillions of spores to be released.

ANATOMY OF A MUSHROOM

The anatomy of the fly agaric, *Amanita muscaria*, is typical of members of the Basidiomycota.

1. The **CAP** (pileus) is at the top of the mushroom structure. It looks like an umbrella and provides a protective covering for the gills.

2. The **GILLS** (lamellae) hold spores in their thin, papery blades under the cap.

3. The **UNIVERSAL VEIL** is a layer of tissue membrane that envelops an immature mushroom to protect it during its development. This breaks apart when a mushroom matures, leaving remnants on the cap and sometimes forming a volva at the bottom of the stipe.

4. The **PARTIAL VEIL** is a tissue membrane that is much thinner than the universal veil. It covers and protects the gills from the edge of the cap to the stipe as the mushroom is developing. Once the mushroom is mature, the partial veil breaks away to expose the gills, leaving a ring (annulus or skirt) around the upper part of the stipe.

5. The **STIPE**, sometimes called a stem, supports the cap so that the gills can easily release their spores into the wind. At the base of the stipe is the volva left by the universal veil.

6. **SPORES** are a fungus's microscopic reproductive cells. Many fungi release their spores into the air to restart the fungal life cycle far from the parent mushroom.

7. **MYCELIA** are white, thread-like filaments made up of hyphae that form the vegetative part of a fungus.

1.5

MYCELIUM INTELLIGENCE

Forests, grasslands and woodlands are not landscapes of individual trees competing with one another for survival. These ecosystems have been formed over millions of years, and their participants have the ability to negotiate, cooperate, trade, steal and compromise – all in the absence of a brain. Fungi connect them all. Underground mycelia weave the forest into a dynamic network of incredible scale.

Mycorrhizal relationships are far more complex than a single partnership between a fungus and a plant. Hundreds of mycelia can be attached to one plant and, conversely, a mycelium can be attached to hundreds of plants. Mycelium is so fine that a teaspoon of soil can hold hundreds of kilometres of mycelium. Over an area as large as a forest, that is a long information highway for fungi and plants to relay resources and chemical signals across. And they do, constantly.

It is widely accepted that the carbon produced from one tree can be shared with its mycorrhizal partners and other trees. This was discovered by Dr Suzanne Simard, an ecologist and professor at the University of British Columbia, and published in a 1997 paper in *Nature*.[14] She called it the 'wood wide web'.

Today, the phrase 'wood wide web' is used to describe the mycelial highways that facilitate the transfer, because they function like the forest's organic internet. Plants within the network can transfer sugars, hormones, stress signals and carbon. Simard mapped the mycorrhizal networks in numerous forests and found that they were structured in a similar way to neural networks in the brain and node links within the internet. The oldest and largest trees had the most mycorrhizal connections – Simard calls them 'mother trees'. She believes that they are social creatures that support the rest of the network by feeding seedlings and injured or shaded trees, warning others of attacks and transferring their nutrients to neighbouring plants before they die.

A close-up of the mycelial network, made up of an interconnected matrix of hyphae. Mycelium is one cell wall thick and about 0.2–5 microns wide (for comparison, a strand of human hair is around 50 microns wide).

Not all scientists agree that fungi and trees operate from a place of altruism and cooperation. Dr Toby Kiers, a professor of evolutionary biology at Vrije Universiteit in Amsterdam, believes that 'both parties may benefit, but they also constantly struggle to maximise their individual pay-off'.[15] Using market economics as a metaphor, Kiers's team published studies showing that plants and fungi trade under free market principles.[16] In some experiments, fungi hoarded nutrients in their mycelium to decrease the supply. With increased demand from plants, fungi inflated the price for the same nutrients. The team also found that some species of plants hijack mycelial networks and steal energy for survival. This is the case for the ghost plant (*Monotropa uniflora*), a translucent white plant that no longer produces green leaves to photosynthesise. In Kiers's work, capitalist fungi and plants display similarities to humans in their ability to manipulate the supply and demand of the forest market.[17]

Fungi don't possess intelligence in the form of a brain or centralised nervous system. They have a series of nerve nets distributed throughout the mycelium through which chemicals can travel, similar to our neural transmitters. These chemical signals trigger responses that are programmed into their DNA, an intelligence that, in many cases, rivals the human brain in its degree of intricacy, complexity and connections. Fungi are sentient without thought; sophisticated without cognition. It's a fungal world, we are just living in it.

Underground mycelia weave the forest into a dynamic network of incredible scale.

Classifying fungi

DR TOM MAY

Mushrooms, puffballs, polypores, coral fungi, moulds, mildews, yeasts, rust fungi and vegetable caterpillars – these are just some of the forms exhibited by members of the kingdom of fungi. To organise the astonishing diversity of fungi, and indeed all life, a classification is used. The basic unit of biological classification is the species. The name of a species is a binomial that pairs the name of a genus and a species epithet. For the fly agaric, *Amanita muscaria*, the genus name is *Amanita* and the species epithet is *muscaria*.

Each species is placed in a hierarchy of taxa, from genus through the ranks of family, order, class, phylum and kingdom. Further ranks can be interpolated, using the prefixes 'sub' and 'super': subclass sits between order and class, and superclass sits between class and kingdom.

A. muscaria belongs to the family *Amanitaceae* within the order *Agaricales*, which sits within the class *Agaricomycetes* of the phylum *Basidiomycota*.

Saccharomyces cerevisiae (also known as brewer's yeast) belongs to family *Saccharomycetaceae* within order *Saccharomycetales* within class *Saccharomycetes* within phylum *Ascomycota*.

CATEGORY	FLY AGARIC	BREWER'S YEAST
Kingdom	*Fungi*	*Fungi*
Phylum	*Basidiomycota*	*Ascomycota*
Class	*Agaricomycetes*	*Saccharomycetes*
Order	*Agaricales*	*Saccharomycetales*
Family	*Amanitaceae*	*Saccharomycetaceae*
Genus	*Amanita*	*Saccharomyces*
Species epithet	*muscaria*	*cerevisiae*

Think of the classification of life as a gigantic library where each book is equivalent to a species. The library is arranged over separate floors (the phyla), and on each floor there are separate rooms (the classes) and so on, until you come to individual shelves (the genera), where the species are lined up. Some shelves have many species, some have few, but each species has a place in the overall system.

Biological classification aims to organise life to reflect evolutionary relationships. Species within the same genus are more closely related than those placed in different genera. Looking at this a different way, the common ancestor in evolution is more recent for all the species within a genus, compared to all the species in a family, and even further back in evolutionary time for all the species within an order, and so on. Mycologists expect that there are several million species of fungi, so placing them all within a classification is essential to keep track of relationships.

Dr Tom May is a principal research scientist in mycology at the Royal Botanic Gardens Victoria, where he has followed his passion for fungi over three decades of collecting, research, publishing and outreach. He has been active in local and international conservation, nomenclature and taxonomic initiatives, including current roles with the International Commission on the Taxonomy of Fungi and the Nomenclature Committee for Fungi. Tom's role in founding the citizen science organisation Fungimap was recognised with an Australian Natural History Medallion in 2014.

Placing species in a classification also makes identification easier. Knowing which genus a mushroom belongs to narrows down the number of species that it might match. Classifications also have predictive power. When you want a book on a particular topic, going to one shelf in the library is much quicker than searching the whole library. If you have discovered a species with interesting properties, such as the production of a certain chemical of use in industry or medicine, other closely related species are a good place to look for subtle variations that might have enhanced versions of the property of interest.

When building a classification, there are two things happening. Firstly, the species need to be discovered, delimited and named, and secondly, the species must be placed in the overall classification. Species delimitation involves carefully comparing a novel, or new, species against its closest relatives to determine that it is different, and essentially that it represents an independent evolutionary unit.

The features used for delimiting species and placing them in an overall classification are similar, but the tool kit of traits has evolved over time. Initially, visible characteristics of fungi were used that could be readily seen by the naked eye, such as shape, size and colour. From the middle of the 19th century, microscopic characteristics that are visible under high magnification, such as the surface features of spores, were integrated. From around the beginning of the 21st century, DNA sequence information became of prime importance for fungal classification. Other characteristics used include chemical and ecological features, such as whether they are hosts for symbiotic organisms. Ideally, information from a number of lines of evidence is used.

For DNA, initially, small sections were used to reconstruct the tree of life. Increasingly, information from across the genome (the sum total of DNA in an organism) is analysed. DNA is the blueprint for a species. It contains the instructions for the entire organism. While the outward appearance of two different fungi might seem similar, the microscopic features, such as spores, may vary considerably, and features at the cellular level may also be fundamentally different. By analysing DNA sequences, we go straight to the underlying instructions. This is particularly useful because parallel evolution is common in fungi, meaning that quite unrelated species may share features. For example, fungi with gills (mushrooms) have evolved independently in various orders in the *Fungi* kingdom.

Integration of information from DNA sequences continues to rewrite the classification of fungi at all levels. Phyla are proliferating, with more than a dozen included in some recent classifications, and there are constant rearrangements at all ranks. This rearrangement is both frustrating and exciting! Names continue to change, even for common and long-known species, but the flip side is that these name changes reflect our improved understanding of the tree of life and how the features of fungi have evolved.

Many of the estimated several million species of fungi are currently known only from small fragments of DNA that have been recovered from soil or water samples. The DNA is different enough that it certainly represents novel species, and perhaps even novel phyla, but there is no physical specimen. The existence of these 'dark taxa' of fungi, named by allusion to the 'dark matter' of the universe, is a challenge for mycologists, because the current rules of naming require a physical specimen. There is lively debate at the moment about how to deal with naming organisms known only from DNA sequences. Whichever solutions emerge, it is vital to collect and classify the dark taxa so that they can be effectively conserved and utilised. One thing is sure – mycologists will not run out of work for quite a while!

FOOD

The most widely cultivated mushroom in the world is *Agaricus bisporus*. White button, chestnut, portobello, Swiss brown, cremini and champignon mushrooms are all the same species – *Agaricus bisporus*. They are simply harvested at different stages of maturity and marketed using different names.

FUNGI CAN FEED US

Humans are the only animals that have learned to cook. Cooking uses chemical reactions to transform raw ingredients into a rich array of sensory pleasures. For most people, food is not just for sustenance. Meals form the fabric of our society, and it is around a table that we celebrate and share our cultures. The sizzle of a steak, a steamy bowl of miso soup, and seven-course degustations: these are all achievements of culture. The study of food illuminates much about the society of a region, revealing its circumstances, values and beliefs.

Fungi have fed us from the dawn of time as a source of healthy, nutrient-dense food. Communities around the world have long foraged for mushrooms to use as food or to support themselves through sales of wild harvests. More recently, people have started cultivating mushrooms in their homes, using organic waste or beginner-friendly grow kits. We have become more aware of our health and the role mushrooms can play in boosting our immune system. This self-sufficiency also improves the resilience of communities during social and economic crises.

Microscopic fungi are also hard at work making our food more nutritious and flavourful. These fungi cannot be seen with the naked eye, but they play a pivotal role in the production and fermentation of some of our favourite foods, such as cheese, wine and beer.[1]

2.1
MICROFUNGI IN FOOD PRODUCTION

Microfungi are microscopic metabolic wizards. Their ability to consume and transform a range of organic matter is the foundation for many food production processes such as fermentation and modern biotechnology. Microfungi used in food production include single-celled yeasts and multicellular moulds.

FERMENTATION

Around ten thousand years ago, hunter-gatherers pursued agriculture and learned to domesticate grains, plants and animals. They were humanity's earliest farmers. The next challenge was food preservation: how do you preserve harvests to ensure a stable food supply through winters and droughts? Fermentation was the key to tasty and nutritious food preservation, and the key to fermentation is fungi.

Fermentation is the decomposition of organic matter using microorganisms such as yeast, mould or bacteria. Essentially, letting food go 'bad' in a controlled way. Many of our favourite foods are made from the controlled decay of ingredients using fungi as the lead decomposer. Milk transforms into cheese, yeasted dough rises and becomes bread, and grain decomposes into alcohol. As a society, we view death as a tragedy, yet death is what gives life its richness. Fungi decay food, but they also breathe a new life of rich flavour into it: think kimchi, salami and ketchup. Cultures of fungi and bacteria create cultures of food. Fermentation makes food more digestible, healthy and delicious. It can even increase the vitamins and minerals in food and make them more available for absorption.

Alcohol was one of the very first ferments. It uses yeast to initiate decomposition. This is a natural process that has been used to make alcohol long before the science behind it was understood.

All the hunter-gatherers needed for their earliest brews was some ripe fruit filled with sugars, wild yeasts (whose spores are omnipresent in the air) and a bit of time. Yeasts eat the sugars (glucose) and produce alcohol and gas (ethanol and carbon dioxide) as by-products. For those scientifically inclined, the chemical equation goes like this:

$$C_6H_{12}O_6 \text{ (glucose)} \rightarrow 2\ C_2H_5OH \text{ (ethanol)} + 2\ CO \text{ (carbon dioxide)}$$

The process completely transforms the fruits. The alcoholic liquid turns bubbly, and the smell and flavour of the resulting concoction change dramatically. Considering the euphoric state alcohol brings, it's no surprise the ancient Greeks accredited this miraculous transformation to a god: Dionysus, the god of fertility and wine (and by extension, fermentation).

Alongside alcohol, essential nutrients such as vitamins B1, B2 and B3 are produced from this process, which benefits energy levels, improves brain function and promotes cell health. It's no wonder beer was a dietary staple for early civilisations. Today, kombucha, yoghurt and kefir fill that space for us.

Evidence discovered in Jiahu, China, confirms that farmers intentionally fermented a mix of rice, fruit and honey in clay jars 9000 years ago.[2] Similarly, wine was fermented as early as 7400 years ago in Persia (now Iran), following the domestication of grapes.[3] Alcoholic beverages were invented many times and on nearly every continent.

How could we not consider fungi nature's alchemists? Almost every plant with some sugar or starch was used for fermentation into alcohol: apples, tree sap, cocoa, corn, berries, rice, sweet potatoes and pineapples, to name a few. The desire for alcohol by early civilisations was as universal as fungal spores in the atmosphere.

Since yeasts are microscopic, their discovery was not as simple as identifying ingredients in a recipe. It wasn't until the invention of the microscope in the 19th century that scientists could observe the fermentation process and, finally, the microorganisms responsible. Before this, nobody knew that the alcohol and carbon dioxide produced during fermentation was due to a tiny, single-celled fungus called yeast. We could finally thank the strains of yeasts, particularly the ubiquitous

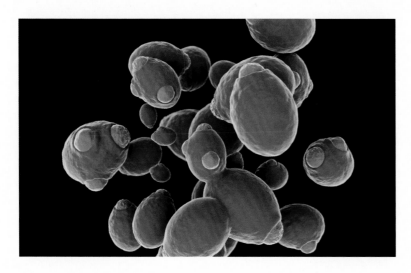

The famous yeast *Saccharomyces cerevisiae*, or brewer's yeast. The smaller yellow spheres are clones, which eventually split off as new yeast cells.

brewer's yeast (*Saccharomyces cerevisiae*), that toiled, broke down the sugars and produced our favourite drinks.

As for bread, before fermentation there was only flatbread. Bread made with yeast rises due to the carbon dioxide from the fermentation process creating air pockets in the dough. These air pockets hold the aroma compounds of bread and trigger olfactory pleasures when eaten. This could explain why bread has been central to the formation of early human civilisations and has remained a staple across various cultures since.

We also need to appreciate the multicellular counterparts of yeast: moulds. Thank you, *Penicillium roqueforti*, for growing throughout blue roquefort cheese to produce the intense flavour and aroma. Thank you, *P. camemberti*, for growing on the surface of brie and camembert cheeses to produce the soft, creamy texture. *Penicillium*, you exceptionally diverse genus of fungi, thank you for antibiotics and contraceptive pills as well.

Moulds also made their mark on the East. In particular, the genus *Aspergillus* plays a starring role in the evolution of food in Japan and China, growing its white, fuzzy colonies in warm, humid and rainy climates. Before you recoil in disgust, think of miso, soy sauce, rice wine and rice vinegar: all these cornerstone flavour bombs in Asian cuisines are fermented using moulds.

Let's take soy sauce as an example – no East Asian pantry is complete without a bottle of this umami goodness. Hailing from 200 CE China, soy sauce (or *shoyu* in Japan) is produced by mixing soybeans, wheat or barley flour as the substrate and inoculating it with *Aspergillus oryzae* (commonly known as *koji* in Japan or *qu* in China). As the mould begins to consume the substrate, salt and water are added and the mixture is left to ferment for six months. The liquid that is subsequently pressed out from the substrate is soy sauce.

In traditional soy sauce production, large vats are filled with the chosen substrate and inoculated with mould. The vats are left out in the sun to let nature complete the process – heat and time allow the mould to ferment the substrate. One of the amino acids released in this process is glutamate. When glutamate is combined with sodium, it creates monosodium glutamate (MSG), a delicious flavour compound found in most condiments. MSG imparts an intensely pleasurable and salty flavour called umami, the fifth taste that stimulates our taste buds alongside sweet, salty, sour and bitter. Today's production of high-quality soy sauce hasn't changed much, except that the sun is swapped with a temperature-controlled room.

In the West, ferments using the *Aspergillus* genus are uncommon. In the 18th century, Europeans made soy sauce using portobello mushrooms (*Agaricus bisporus*) instead, due to their natural glutamate content and accessibility. A large variety of meats such as salami and ham were cured through artisanal methods of salting, drying and fermenting using moulds and bacteria. Regulated for time, temperature and acidity, cured sausages are usually inoculated with *Penicillium* fungi, particularly *P. nalgiovense*, which protects the meat from competing microorganisms and brings out strong flavours.

It's time for an image makeover for moulds. They aren't just menaces in apartments and on rotting fruit. Moulds are the key to delicious and nutritious ferments in diets around the world.

BELOW

Penicillium roqueforti, a mould used to make blue cheeses.

BIOTECHNOLOGY

From this ancient relationship between fungi and humans, fermentation set the stage for modern-day biotechnology. We learned to industrialise the power of fungi to create commercially significant products used in everyday life: processed foods, medicines, washing detergents, textiles and, more recently, alcohol production for biofuels. Our modern lifestyle is dependent on fungal biotechnology for far more than we are aware of – even chocolate and coffee undergo industrial fermentation using yeast.

The 1920s ushered in a new era of food production after the discovery of the mould *Aspergillus niger*. This fungus could create large amounts of citric acid through controlled fermentation, using different types of sugars as the substrate. Companies use citric acid to extend the shelf life of food products and to improve the flavour and texture. The term 'biotechnology' was coined by the agricultural engineer Károly Ereky after Pfizer commercialised citric acid for the food, drug and chemical industries.[4] A century later, citric acid is still produced using *A. niger*. One humble fungus underpins a multibillion-dollar market that flavours and preserves foods such as sweets and soft drinks.[5] Other fungal fermentations have spawned an extensive list of valuable compounds for food processing, such as preservatives, proteins, vitamins, fats and oils.

Fungi's natural capability to secrete enzymes and metabolise a diverse range of organic matter drives innovations in biotechnology. Enzymes speed up chemical reaction, which is important in industrial production. They also allow processes that would otherwise require high temperatures and harsh chemicals to be undertaken in far milder and more energy-efficient conditions using enzymes. Nearly two-thirds of the enzymes used across all industrial manufacturing are derived from fungi. Considering that there are up to 6 million species of fungi, most of which are undiscovered, there may well be methods for making industrial processes better for the planet that are just beyond our reach.

BELOW

Radial growth of the mould *Aspergillus niger*.

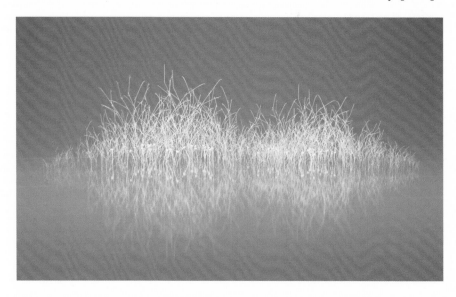

2.2

MYCELIUM AS FOOD

Fermentation has come a long way and delivered many surprising results. One of these delightful accidents is tempeh, a staple in plant-based diets since its origin in Indonesia in the early 1800s.[6] Historians believe that the creation of tempeh was accidental. It may have been discovered while attempting to preserve soybeans from the heat overnight.[7] During this process, the soybeans caught the spores of a fungus, which triggered a fermentation process and formed the compact cake we now know as tempeh. The fungus responsible is *Rhizopus oligosporus*. It eats through the substrate, usually soybeans or other legumes, and binds the substrate together in a web of cottony mycelium that is 100 per cent edible and filled with protein, minerals and vitamins.

David Zilber, the former head of fermentation at the restaurant Noma, has taken tempeh to a new level. As part of the plant-based movement, chefs worldwide have tried to replicate the beef patty in burgers using meat alternatives. Zilber created a tempeh patty made from quinoa. The quinoa grains were colonised by mycelium and left to ferment in the open air to reduce the moisture content, retaining only enough to keep the patty juicy when it was cooked. Finally, the patty was seared with a glaze of yeast garum and fava bean shoyu, which are Noma-made flavour fermentations using fungi. This burger has been hailed as 'the best veggie burger' by taste testers. Zilber commented that 'three fungi and a grain prove that, just maybe, with a little know-how, good cooking can help save, and feed, a world in need of healing'.[8]

What makes tempeh nutritious, and why has it become a wonder food? Not only does tempeh contain some of the fundamentals of our diet – proteins, carbohydrates and fats from the soybeans – the mycelium also offers benefits similar to mushrooms. It is rich with all nine amino acids; packed with fibre, vitamins and minerals; low in calories; and free from cholesterol. So it is not only mushrooms that can be eaten, but mycelium too. The best part is that the mycelium of some fungi has a texture very similar to meat, which is why it's become a popular option on the plates of plant-based eaters.

In Winston Churchill's 1931 essay 'Fifty Years Hence', he predicted 'new strains of microbes will be developed and made to do a great deal of our chemistry for us', concluding that 'synthetic food will, of course, also be used in the future'.[9] Churchill was spot on. In 1985, Marlow Foods launched Quorn, a product line of vegetarian pies made with fungal mycelium branded as 'mycoprotein'. Its commercial success is owed to the fungus *Fusarium venenatum*, which could convert starch into high levels of protein quickly. Their patent for this production process lapsed in 2010, allowing other players to enter the mycoprotein space. But Quorn continues to be ubiquitous in supermarkets today, offering an ever-growing range of meat-free and soy-free versions of poultry, beef and fish.

Ecovative is a biotechnology company founded by Eben Bayer and Gavin McIntyre in 2007 that is taking the next logical step and harnessing fungi to fabricate mycelium materials for packaging, textiles and meat alternatives. Their latest brainchild is Atlast Food, where they manipulate mycelium into synthetic meat. Regulation of temperature, airflow, carbon dioxide supply and humidity encourages the mycelium's fibrous tissues to grow into the shape of various cuts of meat. This complex process is a form of fermentation that allows mycelium to develop different textures, strengths and fibre compositions, resembling animal meat within ten days.

The hope is that mycelium-based meat will reduce the burden of animal agriculture on the planet. Atlast Food's production facility is made up of a vertical farming infrastructure that requires ten times less land and processes less carbon dioxide than traditional meat production does. Atlast Food's first product, mycelium bacon, also uses 100 times less water than conventional pork production.

Advances in biotechnology have allowed this industry to find viable solutions to create sustainable food sources for the future. Why should we take from nature when we can artificially grow food using fewer resources and causing less harm? For everyone who is waiting, Bayer says global availability is possible within three years.[10] The mycelium revolution is upon us.

SIX COMMON SUPERMARKET VARIETIES

OYSTER
Pleurotus ostreatus

ENOKI
Flammulina velutipes

BUTTON
Agaricus bisporus

SHIITAKE
Lentinula edodes

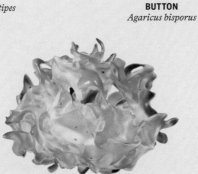

SNOW FUNGUS
Tremella fuciformis

BROWN BEECH
Hypsizygus tessellatus

2.3

MACROFUNGI AS FOOD

Macrofungi means 'big fungi' in Latin. The name refers to the sporing body of this kind of fungi, also known as the mushroom. These sporing bodies have evolved to not only be visually unique but also nutritious and tasty. Not long ago, the public image of mushrooms in Western culture was limited to sliced ornaments on pizza, a sauce for a pub steak or a canned soup. Their nutritional value was rarely understood, and they were disregarded as a peripheral member of the vegetable grocery aisle.

We now know better. Neither a plant nor an animal product, mushrooms are a distinct food kingdom. Every mushroom has a different nutritional profile, but they are all between 20 and 40 per cent protein on a dry weight basis and filled with nutrients. The humble mushroom is actually an ideal food as it is cholesterol-free, sodium-free, gluten-free, low in fats, sugars and calories and an excellent source of essential vitamins and minerals.[11] They are also complex carbohydrates and rich in dietary fibre, like plants – a true all-rounder.

Gourmet mushrooms carry more distinct and complex flavours than the common button mushroom. Once reserved for restaurant chefs, more varieties are increasingly available for the home cook. In particular, the oyster (*Pleurotus ostreatus*), king oyster (*Pleurotus eryngii*), brown beech (*Hypsizygus tesselatus*) and enoki (*Flammulina velutipes*) varieties have become more widely cultivated and sold.

It's unlikely you'll see gastronomical favourites such as white truffles (*Tuber magnatum*) and matsutake (*Tricholoma matsutake*) hit the shelves, as they are yet to be commercially cultivated. This is not for lack of trying – farmers and scientists have attempted this potentially lucrative endeavour for decades with little success. These mushrooms form mycorrhizal relationships with select species of trees, so it is difficult to replicate their ideal growth environment.

This unpredictability and scarcity has driven the cost of the most expensive mushrooms in the world, the exquisite European white truffles, to between US$4500 and US$7500 per kilogram. Depletion of their natural habitat from climate change, droughts, deforestation and unsustainable harvesting also shrinks their availability.

Affordable wild species include yellow morels (*Morchella esculenta*) and chanterelles (*Cantharellus cibarius*), which are foraged and sold at local farmers' markets when in season in the northern hemisphere.

Giving local a new meaning, Smallhold is an American urban farm cross-technology company that stations mini mushroom farms inside grocery stores. Their mission is to solve the under-representation of exotic varieties by distributing climate-controlled machines in which mushrooms grow. Lucky residents in Texas and New York can source lion's mane (*Hericium erinaceus*) and maitake (*Grifola frondosa*) from these local dispensaries without sacrificing freshness and taste.

For an expedition into exotic dried mushroom varieties, visit an Asian supermarket for vacuum-sealed bags of wood ear (*Auricularia auricula-judae*), snow fungus (*Tremella fuciformis*) or shiitake (*Lentinula edodes*). All you need to do when you get them back home is rehydrate them in water and watch them triple in size, then add them to soups or stir-fries or use them as a meat substitute in any dish.

The plant-based food movement is well underway, and mushrooms have taken centre stage to replace meat in plant-based versions of classic dishes like burgers, steaks, tacos and dumplings. Mushrooms are made up of thick collections of hyphae; the more entangled the hyphae, the more chewy, tough and meaty the texture.

Fable, an Australian company, uses mushrooms as the main ingredient of plant-based meat alternatives. One of their recipes uses shiitake and a shortlist of all-natural ingredients to produce what resembles pulled pork, braised beef and beef brisket – all with minimal processing. They source shiitake from farms that would otherwise throw away imperfect mushrooms that do not meet supermarket standards. Fable solves two major problems: avoidable food wastage and the excessive processing involved in some vegan alternatives. Their products are available in Australian supermarkets and are even served in Heston Blumenthal's restaurants.

Veganuary, a UK non-profit organisation, runs a 31-day pledge each January to raise awareness about the benefits of a plant-based diet. Large food chains roll out vegan menu items each year using fungi. Popular UK chain Wagamama launched menu items such as chilli 'squid' made from oyster mushrooms and 'no-duck' donburi made from shiitake and seitan (protein-rich wheat gluten).

Taste and texture aside, the nutritional properties of mushrooms are elevating them into the ranks of superfoods, alongside kale, quinoa and spirulina. Most meat replacements, such as vegan sausages, while tasty and useful as transitional foods, have yet to be refined into truly healthy options.

One 100 gram serving of raw button mushrooms contains 31 calories, 4 grams of carbohydrates, 3 grams of protein and almost no fat. As for micronutrients, the same serving contains 36 per cent of the recommended dietary intake (RDI) of selenium, which supports the immune system and prevents cell damage;[12] 14 per cent RDI of potassium, which helps regulate the nervous system;[13] 13 per cent RDI of phosphorus, needed for cell and bone health;[14] 9 per cent RDI of folate, necessary to maintain healthy red blood cells;[15] and 6 per cent RDI of zinc, which also supports the immune system.[16]

Mushrooms offer nutrients that are predominantly found in meat products, such as amino acids, iron, vitamin B12 and vitamin D. For those following a plant-based diet, mushrooms provide nutrients that would otherwise need to be supplemented.

Another building block of human nutrition is protein. Edible mushrooms provide a high-quality source of protein that is produced with fewer inputs and energy than animal protein, while containing all nine essential amino acids for complete protein intake. Oyster and button mushrooms have the highest levels of protein at about 7 per cent RDI per 100 grams.

Mushrooms are in vogue, thanks to the growing consciousness of consumers seeking healthy meat alternatives that are kinder to the planet and the animals we share it with. All around the world, people are more concerned with what we are eating, how it is produced and marketed. No matter how you look at it, the meat industry is resource-intensive. It uses dramatically more water and land than plant foods and mushroom production do, while emitting nearly a fifth of all global greenhouse gases – more than the whole transportation industry combined.[17] If and when we do choose to eat meat, we should recognise the journey it has taken to reach us.

The culture of food is shifting towards organic food, local farmers' markets and artisanal food. Businesses around the world are starting to reflect the environmental costs of food in their pricing. Soon the carbon footprint of food will be displayed alongside its caloric information to educate shoppers on the ecological impact of their meals. Diversify your palette with more mushrooms to ease the pressure on the meat and dairy industry, and nourish your body while you're at it. Genuine shifts come from individual choices – vote with your fork, chopsticks or fingers for a healthier you and a healthier planet. Mushrooms are entering the mainstream.

The humble mushroom is actually an ideal food as it is cholesterol-free, sodium free, gluten-free, low in fats, sugars and calories and an excellent source of essential vitamins and minerals.

HOW TO MAKE YOUR OWN VITAMIN D SUPPLEMENTS

Vitamin D is crucial for keeping bones, teeth and muscles healthy. *The Medical Journal of Australia* recommends a vitamin D supplement of at least 400 IU per day if sun exposure is unavailable.[18] For people who do not get enough sunlight, mushrooms are the only natural, non-animal source of vitamin D, provided they have been exposed to the sun.[19] This is easily done at home.

Place your mushrooms on a windowsill, with the gills facing the sunlight, for 15 minutes before you cook them. This simple step turns them into a fantastic source of vitamin D. Just 84 grams of raw, UV-exposed button mushrooms contains over 600 IU of vitamin D and is just as easily absorbed by our bodies as vitamin D supplements.[20]

2.4

FORAGING

In the wake of the COVID-19 pandemic, we unravelled as two crucial building blocks of Maslow's hierarchy of needs, food and safety, were tugged out from beneath us. Panic buying set in as food anxiety fired all survival cylinders in our brains. Modern food systems, already fragile and vulnerable, cracked open.

Thankfully, our ancestors have been here before. Interest in foraging peaks in times of hardship, which reflects the instinctual fear that arises about the future.[21] Consciously or not, we cope with food anxiety by taking back our independence and channelling an ancient ability to find and prepare our own food. We've seen an increase in the number of people finding safety and connection with nature, with urban foragers finding a new taste for wild mushrooms. It's not that foraging will become the predominant way to survive, it's about regaining that empowering feeling of self-sufficiency. Besides, the thrill of the hunt is enough for anyone to keep coming back.

Foraging is a unique activity in today's digital world. It reconnects us with nature. We easily forget that our bodies and instincts were shaped by generations of foragers who lived in harmony with the natural world. The mental and spiritual nourishment we gain from getting out of our heads and into nature cannot be understated. Forests and other natural spaces remind us that there is another universe, and it's just as important (if not more) than the one governed by money, commerce, politics and media.

It's not until you lift the veil and consciously peer in that you realise how diverse and widespread mushrooms are. A trip into the forest reconnects you with the broader ecosystem; it reminds you that you are a part of the web of life, not outside of it. A decaying tree trunk stops being an eyesore and becomes a place of opportunity: bracket fungi – wood decomposers that look like shelves – thrive there, small and frequent. Among dead leaves, on fallen trees, in a field of grass or on cow dung – mushrooms grow everywhere.

Foraging is an unlearning of society. Instead of passively absorbing information, you actively peer into the forest to find fungi. Instead of excess, you take only what you need and leave the rest for others. Instead of dulled senses, you hone the skill of noticing, taking in the mushroom's body, earthy aromas and striking shapes, textures, and colours. Foraging awakens the senses and reconnects us to our body. It's a moving meditation from which you learn about the natural world, feel exhilarated by each encounter and with a bit of luck, maybe even take home some free, delicious and nutritious food. Happy hunting.

PLANNING

Foraging is like navigating life – it's difficult to stick to a strict plan and almost impossible to replicate a previous experience. It's best to go out with an open mind rather than an attachment to the end goal. Foraging is more than the satisfaction of finding a mushroom. It's the process of walking through crunchy leaves, smelling the damp organic scent of the forest and meeting friendly Hungarians with walking sticks and wicker baskets. You will soon realise why fungi are called the hidden kingdom. They are everywhere, but they are elusive. It's almost like playing hide-and-seek, except you're not sure what you're looking for and you don't even know if it exists. But have faith, follow the trees, turn over logs, look under leaf litter, and the process will show you the way. A little bit of planning will go a long way to increase your chances of a healthy yield, so let's get started.

WHERE TO LOOK

Woodlands and grasslands are the two primary environments you'll want to start exploring. Woodlands offer fungi organic matter on the forest floors to feast on and trees to form mycorrhizal relationships with. Depending on the type of mushrooms you are looking for, oak, pine, beech and birch trees are longstanding mycorrhizal partners. Find these and you're one step closer. Grasslands also have an abundance of mushrooms, but they generally contain fewer species than woodlands, due to the decreased diversity of trees and environmental conditions. If these options are too far, try foraging in your gardens or in local parks – mushrooms grow there too.

Australia is a fungal haven. Its history of isolation from other continents and its ever-changing climate and nutrient-dense forests all add to its megadiversity. If you're in New South Wales, head to Oberon. You'll find over 40,000 hectares of pine forests, making it one of the best mushroom-hunting grounds in the country. Here, you'll be able to find the popular edible mushroom saffron milk cap, also known as red pine (*Lactarius deliciosus*). This fungus is said to have landed in Australia by accident, its mycelium attached to the roots of a tree imported from Europe. This carrot-orange mushroom was named *Lactarius deliciosus* in 1821 by English mycologist Samuel Frederick Gray, and rightly so, as *Deliciosus* means 'tasty' in Latin. If you want to find these mushrooms in Oberon, try planning a visit in autumn, between late February to May, when they proliferate.

In the UK, Hampshire's New Forest national park is a ninety-minute train ride from London. Made up of both woodlands and grasslands, it offers its visitors a wide range of flora, fauna and fungi to discover. There are even wild ponies roaming around. The forest is home to over 2500 species of fungi, including the unusual stinkhorn (*Phallus impudicus*), whose smell is more of a stench and whose phallic structure is exactly as it's depicted in field guides. There is also the chicken of the woods (*Laetiporus sulphureus*), which grows on its favourite oak tree, stacked up like shelves. Harvesting is not permitted, so take the time to search, identify and appreciate the fungi, rather than picking. If you're lucky, there may be foraging groups – identifying mushrooms rather than foraging for food – in the area.

Even New York City's Central Park offers up foraging potential. Although mushrooms were not intentionally introduced during the park's creation in the 1850s, the 840-acre park has now registered over 400 mushroom species, proving how far-reaching fungal spores are. Gary Lincoff, a self-taught mycologist known as the pied piper of mushrooms, lived adjacent to the park and organised regular mushroom hunts through the New York Mycological Society. Lincoff was one of the early members of the society, which was reconstituted in 1962 by the avant-garde composer John Cage, who was also a self-taught amateur mycologist and an expert in his own right.

When foraging for mushrooms, it's always helpful to go on your hunt with a local expert who knows specific species and their habitats. If you're looking for a guide, your local mycological society is the best place to start.

WHEN TO LOOK

Mycelium will grow all year round, but only when they are subjected to the right conditions – typically influenced by temperature, light, humidity, and the concentration of carbon dioxide. Certain species are fussier than others, but the average ideal temperatures are between 15 °C and 24 °C, typically during the transition in and out of winter, making autumn and spring great seasons to plan a hunt.

When mycelium absorbs water from its surroundings, an explosive force fills its cells with water and sprouts the mushroom. This is why mushrooms usually appear after rain and during the wettest months of the year. Keep in mind these conditions and they can lead you to the treasure. But take note: peak mushroom season occurs at slightly different times each year, as it is greatly affected by temperature patterns and rainfall.

Among dead leaves, on fallen trees, in a field of grass or on cow dung – mushrooms grow everywhere.

TOOLS, CLOTHING AND EQUIPMENT

Your safety comes first during foraging, and it starts with proper preparation. Forage in pairs or groups, never alone. Wear comfortable clothing, such as long-sleeved shirts and long pants, to discourage bugs and ticks. Wear something bright to ensure you're visible in case you get lost or if there are game hunters in the vicinity. Comfortable hiking boots will help your grip in muddy areas. Once again, resist the temptation to go off on your own, but if the mushrooms call to you, ensure you have a physical map, compass and whistle to help you find your way back. Prepare for the worst and learn to use these items beforehand. Last, but not least, properly identify each mushroom before consuming it and when in doubt, go without.

A region-specific mushroom field guide will always be handy, wherever you are foraging and whatever type of mushroom you are searching for. A good field guide contains detailed information, images of hundreds of specific species and should be portable enough for you to carry along on your hunt. This information will be invaluable for your safety when evaluating poisonous species.

FORAGING TOOLS

Folding knife
IMPORTANT

To cut off the base of the mushroom stipe and trim off any damaged areas.

Soft-bristle brush
IMPORTANT

To clean debris from collected mushrooms. Specialty mushroom knives have both a folding knife and brush.

Picnic-style basket or mesh bag
IMPORTANT

To carry your harvests. The gaps in the basket or bag allow spores to fall from the mushroom. This is an act of conservation to ensure there are many more seasons for hunting.

Localised field guide
RECOMMENDED

To help identify fungi. A guide with images and descriptions will also give clues about poisonous species.

Hand lens
OPTIONAL

This is optional, but it will help you see finer details when identifying fungi.

Pocket-sized notebook and pen
OPTIONAL

To note down details of the fungi you find.

Wax paper or paper bag
OPTIONAL

To separate your mushrooms for protection or identification.

CLOTHING

Wide-brimmed hat

To protect you from the sun, doubling as a mushroom container when in need.

Hiking shoes

To manage rough and wet terrain – they should be comfortable and have closed toes.

Long trousers

To prevent insect and tick bites.

Long-sleeved shirt

To prevent insect and tick bites.

Vest with pockets

To hold your mushroom tools.

Waterproof jacket

To keep you warm and dry in unexpected weather events.

Bright clothes

To help you get noticed if you become lost.

SAFETY
AND OTHER EQUIPMENT

Compass

To navigate. Learn how to use this before you set out.

Whistle

To draw attention to yourself if you get lost.

Water and snacks

To keep you going on long treks.

Photography gear

To snap images of fungi you spot.

Bug spray/repellent

To keep away ticks and mosquitoes.

Sunscreen

To protect you from the sun.

Permit

To allow you to legally pick mushrooms if you are foraging on government land. Check whether one is required with your local parks or forest service.

IDENTIFYING EDIBLE SPECIES

Humans have long played a game of trial and error when it comes to edible foods, and have passed down the knowledge of mushroom identification. This knowledge is now conveniently summarised in field guides and identification charts, complete with clear, detailed images.

The purpose of this book is not to teach you how to identify mushrooms. There are experts who write field guides for that. And though there is no 'right' way to learn, using a field guide will help you familiarise yourself with some of the species available in your area, so that you have a mental image of what you are hunting for.

There is a saying among foragers: 'All mushrooms are edible, but some only once.' Once you've found a mushroom, it's good practice to identify it before harvesting. This will ensure you only handle safe, edible species rather than unnecessarily displacing mushrooms that you can't use. Though toxic to humans, poisonous mushrooms have the same right to exist as any other living organism. They play their part in the wider ecosystem by partnering with trees and providing a source of food to other wildlife.

When you come across a mushroom, consult your guide or even take it home to complete further identification analysis by making spore prints or studying the features under a microscope. Keep repeating this process and the beauty of repetition will help you become an expert.

Do not consume any mushroom unless you are completely certain about its identity. Eating a poisonous mushroom can cause nasty symptoms, and even death. There are many doppelgangers – many poisoning cases come from foragers who hunt in new areas and eat mushroom species that they think are safe because they look similar to varieties they are familiar with in their local area.

To make matters more confusing, mushrooms are sometimes both edible and poisonous, depending on how they are treated. One example is the fly agaric (*Amanita muscaria*), which contains toxins such as ibotenic acid and muscimol. When consumed, it can cause nausea, dizziness and hallucinations. German naturalist Georg von Langsdorff published methods of detoxifying the mushroom by boiling it, which makes the sporing body edible. Put simply: do your research.

Once you've correctly identified your mushrooms, consume them when they are still fresh – typically within a week. Generally, mushrooms should be cooked. However, some popular varieties such as white truffles (*Tuber magnatum*) are perfect to eat raw, as are porcinis (*Boletus edulis*), which Italians serve raw, like a carpaccio.

Another beginner-friendly method for learning about edible species is to join a foraging class or mycological society in your region. You'll be able to learn from an experienced person or fellow mycophile. Having a mentor and guide will be invaluable when you are starting out.

SUSTAINABLE HARVESTING

We also need to think about how we engage and interact with nature when we go into a forest to forage for mushrooms. It's not just what you take from the forest but also what you leave behind and what you give back.

Obviously, leaving litter or organic waste that does not originate from that environment should always be avoided. But to ensure mushrooms grow season after season, it is also important to minimise soil disturbance and increase the dispersion of spores. How to achieve this and how to sustainably harvest mushrooms has caused some debate between foragers.

One method is to cut the sporing body off from the bottom of the stipe, where the mushroom meets the soil or the base of whatever it is growing from. Cutting avoids damaging the mycelium, but it leaves a stump that may rot and infect the area. The alternative method is to twist and pick the sporing body from the mycelium, which stops the stump from forming.

A study in Switzerland revealed that 'long-term and systematic harvesting reduces neither the future yields of fruit bodies nor the species richness of wild forest fungi, irrespective of whether the harvesting technique was picking or cutting'.[22] The study added that trampling on the forest floor, however, does reduce sporing body numbers, as this could damage the mycelium. It's important to avoid digging out the mushroom, as this may have the same damaging effect. Finally, cover the harvested area with organic matter to protect the mycelium network from drying out.

Harvesting etiquette also includes leaving some behind. Don't pick all the mushrooms in an area, unless it's a common species. Whenever there are two mushrooms together, pick one and leave the other. When a mushroom is mature, it loses its flavour and texture – leave these behind so they can continue to release their spores. Another way to make sure your harvested mushrooms can spread their spores across the forest floor is to use a wicker basket or mesh bag.

There are laws prohibiting foraging in public areas and state forests, so research before you go to avoid penalties. Some areas may require a permit or have limits on how much you can collect, and other areas forbid harvesting completely on conservation grounds.

Like anything worthwhile, foraging is a process of learning, trying, recalibrating and trying again. Everything requires time and commitment but the exhilaration of connecting with the kingdom of fungi is like nothing else.

All mushrooms are edible, but some only once.

Slow mushrooming

DR ALISON POULIOT

Slowly, slowly ...

Finding fungi requires slowness, not speed. It's about careful observation, noticing subtle differences. Finding fungi as food is about learning a few species thoroughly, rather than many superficially. This is the notion of slow mushrooming. Like slow food or slow art, it's about care, attention to detail and developing deep knowledge. It's about revisiting traditions and gleaning the latest research. It's about being aware of how our choices and actions affect the world around us.

Fungi are diverse and ubiquitous. There are few places you won't find them. Some are delicious and others are deadly. With interest in foraging for wild food on the rise, learning to accurately identify fungi through slow mushrooming reduces the risk of harming yourself and the environment.

Desirable or deadly?

What is an edible mushroom? Edibility can be a little ambiguous, encompassing everything from 'not poisonous' to 'delicious'. Most species are not likely to be poisonous, but few are appetising. Fungi can be categorised as palatable, edible, edible with caveats, of unconfirmed edibility, suspect, poisonous or deadly poisonous. These categories are useful, but are sometimes imprecise and confusing. Palatability (taste, smell and texture) is subjective. Some species are poisonous but can be rendered edible through treatment. Others can be deadly when even the tiniest quantity is consumed. Most importantly, every forager seeking edible fungi should learn the major toxic look-alike species. After all, it's always better to leave an edible mushroom uneaten than to consume a toxic one.

Dr Alison Pouliot is an ecologist, author and environmental photographer with a focus on fungi. As an ecologist, Alison researches fungi with the objective and analytical tools of science. As someone who daily wanders in the bush, she relates to fungi with an aesthetic and sensory appreciation – a natural history of experience. Alison draws on both science and aesthetics in her efforts to stir a broader public consciousness in the way we understand and relate to the fungi, the forest and all life. Alison is the author of *The Allure of Fungi* and co-author of *Wild Mushrooming: A Guide for Foragers*.

Notice your surrounds

Slow mushrooming begins with understanding the ecological significance of fungi in the context of their environments. Fungi are not isolated entities. They live in close association with other organisms and their habitats. Accurately identifying a mushroom means being able to recognise, for example, the habitat types or plants with which a particular fungus associates.

Many fungi form symbioses (alliances) with other organisms. Those that form between fungi and plants are known as mycorrhizal symbioses. Recognising these relationships has obvious benefits for the forager. Being able to identify the tree genera with which a particular fungus is associated saves the forager from aimless wandering around in the 'wrong' habitat.

Finding fungi also means being able to anticipate the type of substrate (growing medium) in which a particular species grows. Some fungi species live in soil, others in leaf litter or herbivore scats, and many grow in wood. Some grow in living trees, others in fallen wood and some on a particular type of wood or wood of a particular age. There are fungi that are only found in undisturbed habitats, while others favour ruderal (disturbed) environments. Knowing the substrate type where your sought-after species grows helps you identify a fungus by eliminating all those substrate types where it doesn't grow.

Keep in mind that seeing a particular fungus species once doesn't mean you'll be able to recognise it the next time you encounter it. Identification skills develop from direct experience in the field and observing a species in different places and conditions over time, as well as learning to recognise its important diagnostic features. Observing the same species in different habitats and situations allows you to become familiar with the extent of variation that can occur within that species. As a mushroom develops, it can change in shape, form and colour. This is further influenced by where and in what it is growing, as well as its exposure to different weather conditions.

Colour and form

Homo sapiens have evolved to notice colour. It's usually the first thing we notice and comment on when we spot a mushroom. Colour is important for identifying fungi, but it can also be an unreliable feature because it can vary greatly within a species. Colour can change as a natural part of the developmental process or with exposure to wind, sun or rain. Many fungi share similar colours but flipping through a field guide looking for a mushroom of a similar colour will only get you so far. It is important to consider colour and morphology (shape, form, texture and general appearance) in tandem.

Becoming familiar with the different parts of a mushroom (for example, the pileus, lamellae and stipe) and the variations both within and between species is the starting point to finding your way around a mushroom. A small hand lens (x 10 magnification) will help you observe the finer features. Don't forget to also use your sense of touch. Mushrooms vary enormously in texture, and touch reveals details that are not always apparent to the naked eye. Use your fingers to discover whether, for example, a specimen is smooth, velvety, rubbery or buttery.

These are just a few hints to get you started. You could spend a whole day, week or month feeling mushrooms to become familiar with their different textures. You could smell them all to familiarise your nose with the great array of fungus scents and odours. Or you might simply begin by getting to know your local plant species and the fungi that associate with them. All these things take time, and gradually your accumulated observation develops into knowledge. The good news, it seems, is that slow mushrooming is growing fast.

2.5

USING YOUR HARVEST

STORING MUSHROOMS

Mushrooms typically have a water content of over 90 per cent, and release moisture into the atmosphere as they age. So without proper storage, you'll be left with a wet, slimy mess. To keep your foraged goods for as long as possible, place them in a paper bag – it absorbs the moisture more effectively than a plastic bag or airtight container. If you have cling wrap to reuse, puncture some holes in it first. A rule of thumb is that mushrooms will last for a week when refrigerated.

PRESERVING MUSHROOMS

The easiest preservation method is to clean and transfer them into a freezer-safe container. Keep in mind that freezing mushrooms tends to compromise the texture, so it's best to use the thawed mushrooms in soup stocks or chopped into veggie burgers, as opposed to frying them.

    Another great way to preserve mushrooms is to dry them, which allows them to be kept for years in airtight jars. Many people stick to the classic method of drying, which is to put the mushrooms out in the sun for one to two weeks. To test if it's dried properly, the mushroom should snap in half when you bend it, like a crispy chip. You can also use a dehydrator set on a low temperature overnight. Alternatively, an oven on low heat with the door open has the same effect – however, this uses a lot more energy and can turn your kitchen into a sauna. Once your mushrooms are dry, keep them in an airtight jar with minimal humidity. You can even throw in a pack of silica crystals as an extra precaution.

PREPARING MUSHROOMS FOR COOKING

Washing the dirt off your mushrooms with water is a common mistake. Wild mushrooms such as chanterelles, maitake, shiitake and oysters soak up water like a sponge. Washing will only waterlog them and dilute their flavour. Instead, use a brush to remove any dirt or wipe them with a damp paper towel. It's a meditative process.

    When preparing wild mushrooms, refrain from slicing them with a knife. Tear or shred them instead. This allows the mushroom's muscle-like fibres to stay in shape and keep their texture during cooking. Once prepared, fry, grill, roast or simmer as your heart desires.

2.6

CULTIVATING MUSHROOMS

For hundreds of years, cultures around the world have optimised nature's processes to produce a steady supply of mushrooms. Today, large-scale cultivators create employment opportunities for thousands, if not millions, of workers while leaving a smaller carbon footprint than other food industries. Low-tech cultivation techniques are also increasingly available for home cultivators and permaculturalists.

FROM SHIITAKE TO BUTTON

Mushroom cultivation can be traced back to the 13th century, when the first written records of shiitake cultivation were detailed in China's Sung Dynasty.[23] Shiitake *(Lentinula edodes)* is a saprophytic fungus that decomposes wood. Early cultivators chopped down hardwood logs to use as the substrate. They drilled holes on the log surface 5 to 10 centimetres apart, then filled the holes with shiitake mycelium spawn, which is a starter culture of mycelium growing on a substrate. The spawn was covered with wax to protect it from the weather, and the logs were stacked up. The mycelium decomposed the wood over eight to fourteen months until it reached the ends of the logs. This signified a successfully colonised log that was ready to grow mushrooms. If it rained, the logs would grow mushrooms shortly after, otherwise the logs are soaked in water for twenty-four hours to stimulate the growth. After a day or two, tiny mushrooms called pinheads would appear. It only took a few more days for the pinheads to morph into full shiitake mushrooms, which were then harvested. Shiitake continues to be cultivated this way by artisans in Japan and China.

Today, cultivation of most mushroom species has moved indoors, and transitioned from using natural substrates such as wood to agricultural by-products such as corn cobs and peanut hulls. This type of organic matter was previously classified as waste by the agricultural industry, but to mycelium it is a nutritious food source. The move indoors has allowed cultivators to control environmental conditions to make energy, water and land use more efficient. This process is therefore more environmentally friendly and provides a steadier yield and lower prices for consumers compared with outdoor cultivation.

The increase in the global availability of cultivated mushrooms has been reflected in supermarket produce aisles. Consider the many variations of *Agaricus bisporus* – white button, chestnut, portobello, Swiss brown, cremini and champignon mushrooms – its prolific appearance on supermarket shelves is thanks to both its ease of cultivation and its ability to ride in a truck for 2000 km and stay fresh in plastic packaging. The white button was a chance mutation in the 1920s from its original brown form. It was selectively cultivated to ride the positive appeal of white bread at a time when white foods were more attractive to shoppers.[24]

Another reason why *A. bisporus* is the most cultivated mushroom today is because it feeds on a range of organic matter. This is not the case for wild fungi such as chanterelles, morels and truffles, as their food source is from mycorrhizal relationships. They live in association with the roots of plants, so replicating their natural habitat in an industrial setting is more complex.

Mushrooms have only come into the spotlight in recent times, so we are still in the early stages of understanding how to cultivate other fungal species. To date, about twenty mushroom species are cultivated, which is a tiny proportion of the fungi kingdom.

CULTIVATING A RELATIONSHIP WITH MUSHROOMS

Thanks to the increased availability of information, mushroom cultivation is no longer reserved for farms and industrial facilities. You can successfully grow mushrooms at home without any fancy lab equipment. The jars and containers you have lying around your kitchen can be used to get started on your cultivation experiment.

Cultivation focuses on three stages: feeding the mycelium, facilitating its growth while minimising competing organisms, and stimulating the sporing body to form. If you want to dive deeper into cultivation, *The Mushroom Cultivator* by Jeff Chilton and Paul Stamets and *Radical Mycology* by Peter McCoy provide guides on cultivating specific species, working with larger operations and technical nuances as you progress.

Through widespread practice and inquiry, new growing techniques and cultivation set-ups have emerged, blurring the line between science and art. Growing mushrooms can be a surprisingly simple yet empowering process, opening up an avenue to form your own relationship with fungi.

RIGHT

Lentinula edodes growing on oak. They prefer oak logs due to their nutritious wood and ability to retain water. A single log can produce mushrooms for up to five years.

CANTHARELLUS CIBARIUS

COMMON NAME

Chanterelle

FAMILY	Cantharellaceae
GENUS	Cantharellus
SPECIES EPITHET	cibarius

This brilliant and intensely yellow mushroom is a delight to spot among leaf litter. Its abundance makes it one of the most well-known mushrooms in the world. It is so plentiful that companies harvest it commercially for export. *Cibarius* means 'good to eat', and with an annual global market valued at US$1.4 billion, it sure is.

Do not confuse *Cantharellus cibarius* with the jack-o'-lantern (*Omphalotus olearius*), which is a poisonous look-alike mushroom.

HISTORY AND CULTURE

Cantharellus cibarius has a common name in many countries that reflects its early prominence in local cultures. Some of these are *capo gallo* (cock's crest) in Italy, *lisichki* (little fox) in Russia, *dotterpilz* (egg yolk mushroom) in Germany, *ji you jun* (egg yolk or apricot fungus) in China, *canarinhos* (canary mushroom) in Portugal and *jaunette* (little yellow) in France. Most of these names refer to its vibrant yellow colour, which is a result of its betacarotene content.

PROPERTIES

EDIBLE

Yes. Has a fruity apricot aroma, mild taste and firm texture.

NUTRITIONAL PROFILE

A raw 100-gram serving contains 32 calories, composed of 90 % water, 7 % carbohydrate, 1 % protein and less than 1 % fat. Rich in vitamins, providing 30 % RDI of vitamin D. High in minerals such as iron and copper.

MEDICINAL

Yes. Contains compounds with antibacterial, antioxidant, anti-inflammatory and antiviral properties.[25] Used in traditional Chinese medicine to treat eye conditions, lung infections, gut problems and dry skin.

PSYCHOACTIVE

No.

ENVIRONMENTAL REMEDIATION

Yes. They can accumulate toxic metals such as chromium, cadmium and lead in the sporing body.[26]

SPORING BODY CHARACTERISTICS

CAP

2–15 cm wide
Convex then depressed at centre or funnel-shaped, wavy at edges
Yellow to orange-yellow, bruises brown-yellow

GILLS

Orange to yellow
Close or crowded
Run down stipe

STIPE

2–10 cm tall
0.5–3 cm thick
Yellow, bruises brown-yellow
Surface smooth
Texture firm

SPORES

Cream to yellow
Oval

FIELD DESCRIPTION

HABITAT

Grows in mycorrhizal relationships with hardwood or coniferous trees, such as pine, oak and beech.

DISTRIBUTION RANGE

Widespread across Europe, Asia, Africa and North America.

SEASON

Summer and autumn.

LACTARIUS DELICIOSUS

Lactarius deliciosus is a carrot-orange hued mushroom with an elegant vase-shaped sporing body. The body is connected to a short stem that is distinctively pitted, as if pressed with oval stamps. When a piece of the cap or gill is broken off, droplets of saffron-orange milk ooze out. The damaged area then quickly oxidises and turns pistachio-green.

HISTORY AND CULTURE

One of the world's oldest culinary mushrooms, *L. deliciosus* is highly regarded in Russia, the Pyrenees and throughout the Mediterranean. In Russia, it is embedded in the culture and is affectionately known as *rhzhiki*, meaning 'redhead'. Rhzhiki is salted, pickled and served as an appetiser, along with vodka. *L. deliciosus* has even been referenced in early illustrations of fungi made over 2000 years ago in the frescoes of Herculaneum and Pompeii.

COMMON NAMES

Pine mushroom, red pine mushroom, orange milk cap, saffron milk cap, delicious milk cap, *rhzhiki* ('redhead' in Russian)

FAMILY	*Russulaceae*
GENUS	*Lactarius*
SPECIES EPITHET	*deliciosus*

PROPERTIES

EDIBLE

Yes. Has a fruity aroma, nutty but bitter flavour and a meaty texture.

NUTRITIONAL PROFILE

A raw 100-gram serving contains 38 calories, composed of 91 % water, 5 % carbohydrate, 2 % protein and 1 % fat. Rich in vitamins and minerals, such as calcium, iron, manganese, potassium, and phosphorus. High in betacarotene, which gives the mushroom its brilliant colour.

PSYCHOACTIVE

No.

MEDICINAL

Yes. Used traditionally in Russia and France to cure coughs, tuberculosis and asthma. Contains terpenoids and other compounds[27] that have anti-tumour, antioxidant, anti-inflammatory and antiviral properties.[28]

ENVIRONMENTAL REMEDIATION

Yes. Tests show that the mycorrhizal relationship between *L. deliciosus* and *Pinus sylvestris* positively influences the growth of pine trees.[29]

SPORING BODY CHARACTERISTICS

CAP

5–15 cm wide
Convex then depressed at centre or funnel-shaped, wavy at edges
Orange or pink-orange, bruises green
Oozes orange-red milk when snapped

GILLS

Orange to yellow
Close or crowded
Run down stipe

STIPE

2–8 cm tall
1–3 cm thick
Orange
Pitted with orange ovals

SPORES

White or cream
Oval

FIELD DESCRIPTION

HABITAT

Grows in mycorrhizal relationships with coniferous trees, particularly pine trees. Found in groups on sandy soils, grass or pine litter.

DISTRIBUTION RANGE

Widespread in temperate and subtropical areas throughout Europe, Asia, Australia and New Zealand.

SEASON

Summer and autumn.

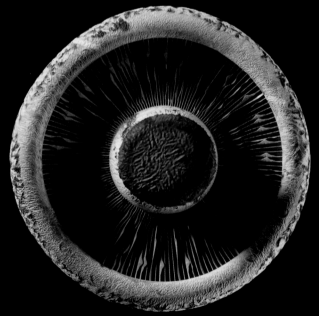

LENTINULA EDODES

COMMON NAMES

Shiitake, black forest mushroom, oak mushroom, *xiang gu* ('fragrant mushroom' in Chinese)

FAMILY	*Omphalotaceae*
GENUS	*Lentinula*
SPECIES EPITHET	*edodes*

Look no further for a dietary staple. *Lentinula edodes* has deep, rich flavours, a velvety texture that can rival meat and it's packed with vitamins, minerals and medicinal compounds. It is widely cultivated and can be found fresh or dried in most Asian supermarkets. The umbrella shaped cap is tan to brown in colour. Its thick caps curl down slightly, with white cracks on the top – the meatier the cap, the more expensive the mushroom. Its strong flavour is concentrated to produce vegan oyster sauce.

HISTORY AND CULTURE

L. edodes is probably the first mushroom to be cultivated by humans. It was mentioned in the written records of Longquan county, China, in 1209. It spread to Japan and farmers refined the cultivation technique by placing logs against trees where it was already growing. Today, it accounts for one-quarter of global mushroom cultivation volumes, second only to the button mushroom.

PROPERTIES

EDIBLE

Yes, very. Prized for its strong aroma and umami flavour.

NUTRITIONAL PROFILE

A raw 100-gram serving contains 34 calories, composed of 90 % water, 7 % carbohydrate, 2 % protein and less than 1 % fat. Rich in vitamins, providing 20 % RDI of B vitamins, and minerals including zinc, iron, manganese and phosphorus.

MEDICINAL

Yes. Rich in medicinal compounds such as lentinans and beta-glucans. Used pharmaceutically in China and Japan as an adjunct therapy for cancer. Used traditionally to lower cholesterol, support the liver, modulate the immune system and lower blood pressure.

PSYCHOACTIVE

No.

ENVIRONMENTAL REMEDIATION

Yes. As a white rot fungus, its mycelium excretes strong chemicals that can break down a range of pollutants. Has been used in environmental clean-ups of xenobiotic matter, such as drugs, cosmetics and industrial chemicals.[30]

SPORING BODY CHARACTERISTICS

CAP

2–25 cm wide
Flat or convex
Light to dark brown
Scaly, white and hairy at edges

GILLS

White
Crowded
Free from stipe

STIPE

5–10 cm tall
0.5–2 cm thick
Light brown
Surface scaly
Texture fibrous

SPORES

White
Oval

FIELD DESCRIPTION

HABITAT

Found grouped together in shaded forest areas. Grows on decaying wood of hardwood trees such as oak, maple, beech, chestnut and hornbeam.

DISTRIBUTION RANGE

Native to Asia, but has been cultivated globally on logs or sawdust.

SEASON

Spring and autumn.

MORCHELLA ESCULENTA

COMMON NAMES

Morel, common morel, yellow morel, sponge morel

FAMILY	*Morchellaceae*
GENUS	*Morchella*
SPECIES EPITHET	*esculenta*

For mushroom foragers, spring is marked by the appearance of *Morchella esculenta*. It has ridges and pits on its honeycomb-textured body, which is usually taller than the stipe. It ranges from brownish-black to yellow and cream, but its distinguishing feature is that it is completely hollow. It is a highly sought-after gourmet mushroom, but attempts at cultivation have been unsuccessful due to its close relationship with the roots of certain trees.

HISTORY AND CULTURE

M. esculenta is the darling of foragers across the US. Some American states even hold annual festivals in its honour. In Michigan, May is 'Morel Month' and a ninety-minute hunt is held to celebrate the arrival of these springtime delicacies. Look in forest habitats that have recently been burned for the best harvest, but beware of the 'false morel' from the *Gyromitra* genus. *Gyromitra* is shaped like a bulging brain and is not hollow.

PROPERTIES

EDIBLE

Yes. Extremely prized for its earthy, nutty flavour and meaty texture.

NUTRITIONAL PROFILE

A raw 100-gram serving contains 31 calories, composed of 91 % water, 5 % carbohydrate, 3 % protein and less than 1 % fat. Rich in iron, copper and vitamin D.

MEDICINAL

Yes. Contains galactomannan, a polysaccharide that can modulate the immune system. Research in animals has shown antioxidant and liver protection properties.[31] Used in traditional Chinese medicine to treat indigestion, improve the function of internal organs and to help dissipate phlegm.

PSYCHOACTIVE

No.

ENVIRONMENTAL REMEDIATION

Yes. Can accumulate metals such as lead and mercury in its sporing body. Can be used as a bioindicator of soil pollution and to remediate contaminated soils.

SPORING BODY CHARACTERISTICS

HEAD

3–11 cm tall
2–6 cm wide
Conical or oval
Yellow, tan or brown
Irregular, spongy ridges and pits

STIPE

1–10 cm tall
1–5 cm thick
Enlarged at base
White to yellow-brown

SPORES

Released from sacs in pits
White to cream
Oval

FIELD DESCRIPTION

HABITAT

Grows in a mycorrhizal relationship with hardwood and coniferous trees, particularly ash, elm and apple. Found in soils in forests, orchards, grasslands, gardens and in recently burned areas.

DISTRIBUTION RANGE

Widespread around the globe in temperate and subtropical regions, particularly in Asia, North Africa, North America and Brazil.

SEASON

Spring.

TRICHOLOMA MATSUTAKE

COMMON NAMES

Matsutake, pine mushroom

FAMILY	*Tricholomataceae*
GENUS	*Tricholoma*
SPECIES EPITHET	*matsutake*

Tricholoma matsutake is a prized delicacy around the world. It is spicy and pungent with a firm texture, and becomes the star of any dish. It is particularly adored in Asian cultures, often given as a gift in important ceremonies to symbolise good fortune, happiness and fertility. *T. matsutake* favours disturbed habitats, growing in the aftermath of forest fires and nuclear disasters. It is believed to be the first living organism to grow after the atomic bomb was dropped on Hiroshima.

HISTORY AND CULTURE

It is not yet possible to cultivate *T. matsutake* due to its complex root association with coniferous trees. Demand is met by harvesting it from forests, but due to climate change, deforestation and inadequate forest management, the supply has dropped sharply. In Japan, harvest peaked at 12,000 tonnes in 1940, but had fallen to less than 100 tonnes per year by 2012. *T. matsutake* is on the global IUCN Red List of Threatened Species. Its scarcity makes it one of the most expensive mushroom species in the world, reaching a price of up to US$1250 per kilogram.

PROPERTIES

EDIBLE

Yes. Considered a gourmet treat due to its powerful flavour and aroma.

NUTRITIONAL PROFILE

A raw 100-gram serving contains 23 calories, composed of 88 % water, 8 % carbohydrate, 2 % protein and less than 1 % fat. Rich in dietary fibre, vitamins and minerals, such as vitamin B, iron, copper and potassium.

MEDICINAL

Yes. Used as a tea in traditional Chinese medicine to modulate the immune system, improve digestion, detoxify the body, protect the liver and clear the skin. A source of polysaccharides, which have been shown to have antibacterial, anti-tumour and immunomodulating properties in mice.[32]

PSYCHOACTIVE

No.

ENVIRONMENTAL REMEDIATION

No.

SPORING BODY CHARACTERISTICS

CAP

5–20 cm wide
Convex or flat
White or tan to dark brown
Brown spots when mature

GILLS

White to tan
Crowded
Attached to stipe

STIPE

5–15 cm tall
1.5–4.5 cm thick
Tapers towards base
White or tan to dark brown
Fibrous upper ring
Firm, dense

SPORES

White
Oval

FIELD DESCRIPTION

HABITAT

Grows in a mycorrhizal relationship with coniferous trees, particularly pine trees. Found in nutrient-poor soils where there is little ground vegetation and a thin litter layer.

DISTRIBUTION RANGE

Grows throughout Asia and northern Europe, and on the Pacific coasts of Canada and North America.

SEASON

Autumn.

Fungi have long been rich sources of medicines for humans. In the last decade, more than one-quarter of Nobel Prizes in Physiology or Medicine have been awarded for discoveries based on the single-celled fungus known as yeast.

FUNGI CAN HEAL US

We are a cosmic wonder. Our bodies are home to some 37 trillion cells, all working in harmony. A natural intelligence propels them forward, day in and day out, for decades, with one singular purpose: to keep us alive. Our bodies are not distinct from the bodies of fungi, microbes, animals or plants. Hundreds of millions of microbes live on and within us – without them, we would not be alive.

When something eventually goes wrong, be it a disease, infection or physical injury, modern medicine is nothing short of a miracle. Medicinal drugs work by changing, destroying or replacing affected cells in the body. It is, without a doubt, all incredibly ingenious. But the decline in health and wellbeing in modern society is the result of a combination of physical, mental, environmental, social and political phenomena. Modern medicine alone cannot fix this.

Increasingly, people are seeking holistic, alternative and complementary therapies. These terms are often used interchangeably and are synonymous with any philosophy or system that aims to improve a person's health and wellbeing outside of the conventional medical system. Many of these therapies, such as herbal medicine, Ayurvedic medicine and traditional Chinese medicine, arose before the scientific discoveries of the 20th century that led to Western medicine.

Because modern and traditional systems of medicine work on different principles, there is a tension between them. What if the differentiation could be as simple as reserving Western medicine for what it is best used in response to – disease, infection and injury – and deploying everything else to manage chronic illnesses or, even better, prevent ourselves from getting sick in the first place?

Remarkably, fungi offer up solutions using both approaches. They are the world's best chemists, containing an array of solutions for prevention and treatment in their medicine cabinet. Many of their traditionally ascribed health benefits are now being affirmed by scientific research, meaning fungi could be powerful supporters of our everyday health and wellness. Meanwhile, biotechnology companies harnessing the power of big data, machine learning and genetic engineering are innovating new methods of drug discovery and production, while pharmaceutical companies are continuing their quest to uncover novel medicines from fungal enzymes.

Hundreds of millions of microbes live on and within us – without them, we would not be alive.

3.1

THE HISTORY OF FUNGI IN MEDICINE

There is a long history of using fungi for their medicinal qualities. Mushrooms have a starring role in traditional cultures due to their dual role as food and medicine. The wisdom of this duality is highlighted in an ancient Chinese proverb, *yao shi tong yuan*, which translates to 'medicine and food share a common origin'. After all, what we consume every day provides the building blocks for our bodies to make and replace cells, and is the source of nutrition for a healthy metabolism and strong immune system.

Traditional Chinese medicine is, in fact, a useful lens through which to consider the potential role of fungi in our lives, as it views the body as a whole system that seeks balance and connection with nature. Qi is the life force or energy flow, and good health reflects a state where qi is circulating the body in equilibrium. Illness and its symptoms reflect an imbalance of energy. Herbal tonics contain medicinal compounds that are believed to replenish qi and bring the body back into balance. This holistic treatment explicitly acknowledges the role of nutrition, exercise and mental health in promoting the body's innate self-healing potential.

Over and above the nutritional value of mushrooms, the medicinal values long ascribed to them are now being validated using modern scientific methods. Microscopic fungi, such as yeasts and moulds, play a big part too. But we have only just started peering into the fungal medicine cabinet to harness their properties to treat human diseases and infections. In the rush to create new medicines, it's worth understanding what ancient cultures have used to great effect for thousands of years.

HISTORY OF MACROFUNGI

The earliest evidence of mushrooms used as medicine comes from Ötzi the Iceman, who lived during the Copper Age. His body was mummified in a glacier in the Austrian Alps 5300 years ago, making him older than the Egyptian pyramids and Stonehenge.[1] For Ötzi, medicine existed in the form of two mushrooms, which he kept in his belt pouch. One was a birch polypore (*Fomitopsis betulina*) with broad antibiotic and anti-parasitic properties. Scans of his body revealed that Ötzi had whipworm. He presumably used the birch polypore to expel parasites from his digestive tract. Ötzi also carried a tinder fungus (*Fomes fomentarius*), which is highly flammable when struck with flint stones. He probably used this for starting fires and sterilising wounds.

Hippocrates, widely regarded as the father of modern medicine, also prescribed tinder fungus to cauterise wounds and treat inflammation in 450 BCE. Alongside Hippocrates, some of the earliest knowledge resides within the pages of the first Chinese materia medica, *Shen Nong Ben Cao Jing* (*The Divine Farmer's Materia Medica*), compiled around 250 BCE. It prescribed fourteen medicinal fungi in its list of 365 herbal medicines and formed the foundation for traditional Chinese medicine.

BELOW

Fomitopsis betulina bursts through tree bark. Its round, rubbery body ages into a corky, tan-grey hemisphere.

There is a species revered throughout Asia that's referred to as the 'mushroom of immortality'. So potent was its effect on strength, vigour and longevity that it was reserved for royalty. Its reddish-brown sporing body grows glossy, flat and fan-shaped from the sides of decaying trees. This fabled mushroom is *Ganoderma lingzhi*, known as *ling zhi* in Chinese, which translates to 'spirit mushroom' or 'divine mushroom'. It was believed that *G. lingzhi* could enhance the body's energy systems, strengthen heart health, improve cognitive function and reverse the effects of ageing.[2] *G. lingzhi* is threaded throughout Chinese culture. It is depicted in art, dances and poems, and even adorns imperial palaces as a symbol for good health, good fortune and immortality. But does science agree?

In the last decade, scientists have published thousands of studies inquiring into *G. lingzhi*'s potential in medicine. Observations made in traditional healing practices are being scientifically validated. *G. lingzhi*'s biologically active properties are attributed to over 300 medicinal compounds. These compounds provide diverse benefits, ranging from immune system modulation to reducing blood pressure, cholesterol and blood sugar levels. Leading the research efforts are China, Japan, Korea and the US, and pharmacological products are being developed for use as antibiotics, antivirals, anti-cancer compounds, blood pressure medication, immunosuppressants, liver protection medications and antioxidants.

It's not just *G. lingzhi* that is bringing hopes of a healthier life. In China alone, 798 species of medicinal fungi were reported in 2018.[3] Other highly praised medicinal mushrooms that promote immune health include snow fungus (*Tremella fuciformis*), cloud ear fungus (*Auricularia polytricha*) and hoelen (*Wolfiporia cocos*).

Medicinal mushrooms have also had a rich history outside of Asia. Chaga (*Inonotus obliquus*), a burnt-looking black mass found growing on the side of trees, has been used in folk medicine in Russia, Siberia and Scandinavia since at least the 13th century. The name 'chaga' is derived from an old Russian word, чага, which means, simply, 'mushroom'. *I. obliquus* was used to treat gastrointestinal disorders, cardiovascular diseases, diabetes and various forms of cancer.[4] In the 1900s, the Soviet Union commissioned a series of studies on the traditional uses of *I. obliquus*, resulting in a licensed medication called Befungin. Befungin is still available in Russia, where it is used to regulate the immune system and treat chronic inflammation, skin conditions, nervous system disorders and early-stage cancers.

Throughout the Middle East, the Mediterranean, North Africa and Western Sahara, desert truffles of genera *Terfezia* and *Tirmania* are revered. The chain of knowledge about the nutritional and medicinal qualities of these desert truffles has been unbroken for 4000 years. Research substantiates many of the traditional uses, concluding that desert truffles contain multiple antioxidants and have antimicrobial properties that are effective against common pathogens.

Before we had any scientific understanding of microbiology, microscopic fungi were used as medicine in the form of mouldy cheese or bread applied directly to the body to treat infections. Today, pharmaceutical companies extract the active compounds from these ancient remedies and millions of patients with life-threatening diseases, from diabetes to cancer, are treated each year with medicines made from microfungi. This contemporary use began with the discovery of penicillin by Alexander Fleming.

By the end of the 1800s, scientists understood that bacteria caused many diseases, and research focused on ways to impede bacterial growth. In those times, even a minor cut could lead to tissue damage, organ failure and death. In 1928, Fleming was studying the bacteria *Staphylococcus* at St Mary's Hospital in London, when spores of the mould *Penicillium notatum* drifted into the lab and landed in a petri dish. Several strokes of good fate facilitated the discovery of penicillin. Fleming was on a two-week vacation when the contamination occurred, he had not packed away his experiments before his departure, and the temperature was warm enough in London to permit the growth of both the bacteria and mould. By the time Fleming returned, he observed that there was a bacteria-free zone around the mould. The mould had not only inhibited the growth of the bacteria, but also produced chemicals that killed the bacteria completely. Fleming named the active chemical 'penicillin' and published his findings in the *British Journal of Experimental Pathology*. Despite its promise, penicillin was difficult to purify and stabilise for clinical trials.

A decade later, Dr Howard Florey and Dr Ernst Chain from Oxford University stumbled upon Fleming's paper. While Fleming is frequently attributed as the sole founder of penicillin, it was Florey and Chain and their colleagues who proved, through a series of gruelling experiments, that penicillin was non-toxic to humans and effective for treating a wide range of infections in the human body. By 1941, in the midst of World War II, they knew that *P. notatum* could never yield enough penicillin to treat people reliably. The search began for high-yielding strains of *Penicillium*.

Fruit and soil samples were submitted from all over the world. In 1943, a laboratory assistant named Mary Hunt sent in a cantaloupe covered in a 'pretty, golden mould'.[5] That mould was *Penicillium chrysogenum* and it yielded a tremendous amount of penicillin – 1000 times as much as *P. notatum* after a few rounds of mutations. The team developed the production process and companies in the US and UK quickly mass-produced penicillin, enabling the first antibiotics to be sold as drugs.

Penicillin was a wonder drug. It cured disease and infections on the front line for the rest of the war. Some historians even credited the Allied victory to the miracle of penicillin. Fleming, Florey and Chain received the 1945 Nobel Prize in Physiology or Medicine for their discoveries.[6] Fleming is widely quoted as saying, 'I did not invent penicillin. Nature did that. I only discovered it by accident.'[7] How many other solutions to human suffering lie undiscovered in nature?

Following penicillin, the pharmaceutical industry roared into a golden age of antibiotic discovery. In 1957, Sandoz, a Swiss pharmaceutical company, created an antibiotics discovery program where employees were encouraged to collect soil samples from trips overseas to uncover new fungal strains. In 1969, a soil fungus called *Tolypocladium inflatum* was discovered in Norway. Research led by Dr Jean-François Borel and Dr Hartmann F Stähelin in the 1970s uncovered *T. inflatum*'s ability to produce cyclosporin, a chemical that can suppress the immune system. At the time, it was known that the immune system defended the body against foreign materials, especially invading germs. However, transplanted organs, no matter how necessary to a patient's survival, were recognised by the body as an invader and rejected. For a transplant to be successful, the immune system needed to be dampened or weakened just enough to prevent rejection of donated organs. Cyclosporin could do just that. This discovery revolutionised the success rate of organ transplants and cyclosporin continues to be the best-selling immunosuppressant today. Cyclosporin is also an effective treatment for psoriasis, severe dermatitis and rheumatoid arthritis.

BELOW
Penicillium moulds produce chemicals that can kill bacteria, as shown by the bacteria-free zone surrounding the mould.

3.2

MEDICINAL BENEFITS

Fungi straddle the worlds of medical science and traditional herbal medicine. As a relatively accessible natural remedy, with few side effects and no known lethal dose, medicinal mushrooms in particular are an increasingly attractive alternative to conventional treatments. Beyond that, fungi permeate the market in the form of powders, capsules, tinctures, drinks and even skincare.

Much more importantly, fungi have the ability to strengthen and regulate our immune systems, protecting us against pathogenic viruses and bacteria. To best understand the medicinal benefits of mushrooms, it is helpful to understand the basics of the immune system.

HUMAN IMMUNE SYSTEM

Of all the wonderful services that the body provides, the immune system is one of the most remarkable. It is our natural defence system. It protects us for most of our lives, despite our limited understanding of how it works.

The immune system acts like border control, identifying and removing any incoming threats. It has physical barriers – the skin is the most important of these, along with hair, mucus and tears. The second line of defence is chemical. Antimicrobial barriers are secreted on the skin, in vaginal discharge and in semen. Our stomach also contains strong acids and large populations of good bacteria that prevent harmful bacteria from colonising the body.

If pathogens manage to get past these defences, the immune system dispatches immune cells from the lymph nodes, thymus, spleen, bone marrow, tonsils and other parts of the body. There are a variety of immune cells, but their uniting purpose is to fight off disease and infections. It is critical that they can tell the difference between healthy and dangerous cells. Most symptoms of illness don't actually originate from microbes but from the activities of our immune system.

Our moods, sense of connection and stress levels have a profound effect on the production of our immune cells. Happy chemicals in our brain, such as serotonin, dopamine, oxytocin and endorphins, strengthen the immune system. But when we are stressed, the body signals to the immune cells to attack healthy cells, which we experience as pain and inflammation. When this happens, we reach for pain relievers, antibiotics and other temporary solutions, but these do not address the underlying issues.

Modern afflictions such as poor nutrition, stress and anxiety can compromise the immune system and leave the body vulnerable to attack. As a result, autoimmune diseases have been on the rise in recent decades. Psoriasis, fibromyalgia and multiple sclerosis are some of the many common autoimmune diseases whose symptoms don't stem from a specific source. Since the underlying issue in many illnesses is the immune system, there is no one cure.

To make things even more complicated, everyone's immune system is different. Treating immune system disorders requires a personalised approach, including changing how people think, feel, eat and live their lives.

WHY DO THEY WORK?

Fungi are constantly engaged in competition with microbes, insects and other animals. Fungi are immobile, so they cannot run away in the face of a threat. Their method of self-defence is to produce an array of complex chemicals that attack and destroy pathogens. Fortunately, fungi have had millions of years to develop their arsenal of chemical survival strategies. And fortunately for us, many of these strategies can be helpful in our fight for health. Many of the compounds that fungi use to fight off pathogens are also effective in our bodies. Perhaps this is because of our evolutionary closeness (we share almost 50 per cent of our DNA and 85 per cent of our RNA with fungi),[8] or perhaps it's simply our long history of co-evolution.

Medicinal fungi are not just antibiotics. The potential for various fungi to produce medically useful molecules has expanded beyond antibacterial and into antifungal, anti-parasitic and antiviral compounds. Other famous medicinal fungi include *Aspergillus terreus*, from which drugs used to lower cholesterol are derived, *Isaria sinclairii*, used to treat multiple sclerosis, and *Claviceps purpurea*, used for migraine relief.

ACTIVE MEDICINAL COMPOUNDS

Although medicinal mushrooms have been used in traditional medicine for thousands of years, identifying beneficial compounds and extracting the active ingredient for mass production is a marker of modern medicine. Compounds found in sporing bodies and mycelium can regulate the immune system, stimulating the body's self-healing potential. Some are even able to attack malignant cells or microbes responsible for infections. But most mushrooms are best suited for preventative care, working with a functioning immune system to increase its defence systems and keep the body free of disease.

The bioactive compounds responsible for these benefits belong to many different chemical groups, including polysaccharides, terpenes, phenolic compounds, alkaloids, peptides, lectins and nucleosides. The two most studied compounds are beta-glucans and triterpenes.

Beta-glucans

Fungal cell walls are made up of chitin, the same structural component that is found in the exoskeletons of crustaceans. Fungal cell walls are rigid, and nested inside them are long chains of complex carbohydrates called polysaccharides. Fungi contain many types of beneficial polysaccharides, but the most important one for humans is a beta-glucan (specifically, the (1-3)(1-6) beta-D-glucan).

Over 20,000 published studies have shown beta-glucans to be the most potent natural immune system–supporting forces ever discovered. They can stimulate weak immune systems and regulate overactive ones. They increase our protection against bacterial, viral and parasitic infections, and offer a host of other synergistic health benefits. Not surprisingly, the species of fungi most revered in traditional medicine are high in beta-glucans. Studies that tested for the level of beta-glucans in mushrooms found the highest beta-glucan content in the sporing bodies of turkey tail (*Trametes versicolor*) and reishi (*Ganoderma lingzhi*).

Our co-evolution with fungi has left our bodies with receptor sites that can receive compounds found in fungi. As we evolved, we capitalised on the fact that fungi, microbes, plants and animals have different processes to produce molecules. We developed pattern recognition receptors that specifically bind to non-human compounds.

Picture a lock and key: when we consume beta-glucans, they fit into specific receptor sites in our immune system and activate our immune cells. Immune cells hunt for pathogens and cancerous cells in the body. By stimulating your immune cells, you stimulate your body's innate healing and defence responses. This means that medicinal mushrooms support nearly all of the body's major systems and allow you to perform at your full potential.[9]

Beta-glucans are found in all fungi, including microscopic yeasts. With more than eighty clinical trials evaluating their biological effects, the question is not *if* beta-glucans will move from supplement to widely accepted drug, but *when*.

Triterpenes

Unlike beta-glucans, which exist in all fungi, terpenes are only found in select species. Terpenes have been studied for their anti-cancer, anti-tumour, antimicrobial and anti-inflammatory properties and their effectiveness in countering neurodegenerative diseases. They work by stimulating our immune cells to attack pathogens that invade the body, while managing the body's response by preventing cells from proliferating unnecessarily. There are different types of terpenes, but triterpenes are the most active and have been well researched.

RIGHT

These diagrams show the make-up of a mushroom, from the superstructure of the sporing body down to the cellular level.

MUSHROOM

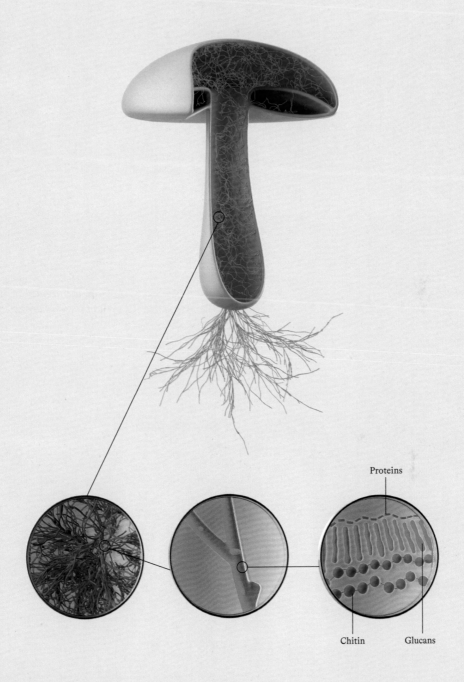

Proteins

Chitin   Glucans

MYCELIUM   **HYPHA**   **CELL WALL**

In 1982, ganoderic acids A and B were the first triterpenes isolated from *G. lingzhi*, or reishi. Since then, more than 300 have been discovered. Triterpenes can inhibit the growth of tumour cells. That doesn't make them first-line drugs for cancer, but Jeff Chilton, the co-author of *The Mushroom Cultivator* and founder of Nammex, an organic medicinal mushroom extract supplier, says, 'I would probably recommend reishi to anybody with a life-threatening illness – take reishi and take a lot of it. Mushrooms potentiate the immune response and as such, should be looked upon as a primary means of disease prevention, not a cure. Eat mushrooms regularly and use supplements as needed.'[10]

Triterpenes can also protect the liver, lower cholesterol, reduce inflammation and play a powerful role in balancing hormones. This makes *G. lingzhi* the ultimate adaptogen. Adaptogens balance hormones and protect the body from psychological and physiological stress, improving the functioning of the immune system. This is important because in our fast-paced lives, stressors are everywhere. Unfortunately, our brains are not very discriminating about what they consider threatening. Modern life – traffic, stress, social anxiety – ignites a body-wide explosion of hormones, commonly known as the fight-or-flight response, just as much as being chased by a predator does. The result? A chronically stressed population, with symptoms manifesting as poor digestion, insomnia, infections, immune disorders and low energy levels. We can help our bodies respond and adapt to this lifestyle by adding adaptogens into our routines. In nature, mushrooms grow in extreme and challenging environments, so they have developed unique methods of survival. When we consume mushrooms, we also receive their adaptogenic qualities. They don't target illnesses directly, but they do strengthen the ecosystems of our bodies.

POTENTIAL OF FUTURE DISCOVERIES

Great progress has been made in understanding the biochemistry of mushrooms, but what we currently know is far from exhaustive. Funding constraints and the expensive process of large, double-blind studies have made the rate of discovery slow. Fortunately, we can safely include mushrooms into our diet and supplement with mushroom nutraceuticals, as we know they are non-toxic and beneficial for our immune system.

Hexagon Bio is one of the new generation of pharmaceutical companies that are applying DNA sequencing, artificial intelligence and data science to fast-track fungal pharmaceutical discoveries. There are currently 5000 fungal genomes that could potentially create new drugs for cancer or infectious diseases. Without Hexagon Bio's ability to analyse vast datasets, the screening process would be extremely slow. Given that there are up to 6 million fungal species available in nature, Hexagon Bio is in a position to harness an unparalleled amount of chemicals from fungi for human medicinal use.

This also highlights the urgency of conserving fungal biodiversity from which future medicines might emerge.

3.3
RESEARCH
AND TESTING

Although there are health claims in the wellness space that undermine the scientific integrity of the research community, the impressive resume of medicinal mushrooms has been compiled through hundreds of animal experiments and test-tube studies. Dozens of clinical trials with human patients have also validated some of these claims, but these studies are not conclusive. Hopefully, medical science will substantiate many traditional uses by ancient cultures.

FUNGI AS A BRAIN ENHANCER

Our brain is our physiological cockpit, firing over 100 billion nerves to provide us with memory, movement, pleasure, pain and a range of complex emotions. As you read this page, your brain is making instantaneous connections across multiple regions to comprehend and create meaning from these words.

The brain is a hungry organ. It uses 20 per cent of our daily energy intake, despite being only 2 per cent of our body weight. Today, our brains work overtime and this manifests as fatigue, brain fog, forgetfulness and emotional instability. Neurological disorders such as Alzheimer's disease are also on the rise – one person develops dementia every three seconds.

Science accepts that the brain reorganises itself and creates new brain cells as we grow and age. Through a process called neurogenesis, new neurons develop in our hippocampus, the part of the brain responsible for long-term memory, learning, and emotional stability. Neurogenesis is regulated by a group of molecules collectively known as the nerve growth factor. Studies suggest that deficiency of the nerve growth factor is linked to Alzheimer's disease.

The white pom pom-shaped mushroom *Hericium erinaceus*, or lion's mane, has captured the attention of scientists due to the discovery of two compounds. One compound, hericenones, was isolated from the sporing body, and another, erinacines, was isolated from the mycelium. In a trial, doses of erinacines were fed to rats through their stomach, which prompted the growth of the nerve growth factor and increased neurotransmitter levels in their brains.

When administered orally in humans, these compounds can easily pass through the blood-brain barrier, suggesting the activation of nerve growth factor synthesis. This is an exciting discovery, as it creates the potential to find treatments for Alzheimer's disease, for which there is currently no cure. While there is no conclusive evidence, a small number of human clinical trials and animal studies continue to indicate that hericenones and erinacines stimulate neural cells, improving cognitive functions through their neuroprotective properties.

In 2008, a double-blind, placebo-controlled human clinical study suggested that *H. erinaceus* can improve mild cognitive impairment.[11] A group of mildly cognitively-impaired men and women aged between fifty and eighty were given 1 gram of dried sporing-body powder orally three times a day for sixteen weeks. When tested, they showed significant improvements in their cognitive functions after eight, twelve, and sixteen weeks compared with a group that was taking a placebo. Four weeks after the supplementation stopped, the participants' scores began to drop. In a follow-up study in 2019, *H. erinaceus* again showed significant improvements to cognitive function.[12] These results were consistent throughout the study, with findings showing that *H. erinaceus* potentially regenerates neural networks in the brain.

Other research substantiates the ability of *H. erinaceus* to promote the nerve growth factor, but it is unclear if the results arise from hericenones.[13] As yet unidentified compounds within the *H. erinaceus* sporing body could be responsible for these positive effects.

Research also shows that a strong immune system can positively impact the brain. When the immune system responds to infections or an injury, it triggers inflammation, which is a signal to defend, repair and heal the body. A 2019 study showed that systemic inflammation led to a cognitive decline in later adulthood. As a healthy body supports a healthy brain, fungi containing beta-glucans with immunomodulatory effects may also benefit the brain and cognition.[14]

FUNGI AS AN ENERGY BOOSTER

We are all looking for a silver bullet when it comes to cultivating lasting energy levels. It's easy to reach for an energy drink, but most brands are filled with sugar and only provide a short energy spike. A promising natural alternative with no side effects is a fungus called cordyceps, which refers to both *Ophiocordyceps sinensis* and *Cordyceps militaris*.

O. sinensis is a parasitic fungus also known as the caterpillar fungus. In China, it is called *dong chong xia cao*, which translates to 'summer grass, winter worm'. It infects ghost moth caterpillars in winter, transforming them into medicinal fungi to be foraged in the summer. The caterpillar hibernates in soils, where spores germinate on its body. The mycelium grows throughout the caterpillar, consuming its insides, and forms a club-shaped sporing body that sprouts through the caterpillar's head. The fungus – and the caterpillar's body – is then consumed by predators, which includes humans.

This may seem bizarre, but *O. sinensis* is one of the most acclaimed medicines in traditional Chinese medicine. It has been used since 620 CE for enhancing energy, vitality, endurance and even sexual libido. It is believed to nourish yin and lift yang – in other words, its adaptogenic properties can balance the body in response to stressors. The persistent demand for *O. sinensis* drives costs up to US$20,000 per kilogram, making it one of the most expensive fungi in the world.

Today, this demand is partly met by sustainably grown variants of *C. militaris*. This newly cultivated species has similar benefits to *O. sinensis* but it's more ecologically sound. Jeff Chilton explains that ten years ago, 'they brought *C. militaris* into cultivation and they're not growing it on any insects. It's just the mushroom, and the price is reasonable. Now it's possible to bring it into the supplement market.'[15] Bringing a mushroom species into commercial cultivation is a rare event, and *C. militaris* has significantly furthered the industry of energy-boosting medicinal mushrooms.

A 2018 double-blind, placebo-controlled study showed that, after three weeks of supplementation with *C. militaris*, participants improved their tolerance to high-intensity aerobic and anaerobic exercise.[16] This included significant improvements in their maximal oxygen uptake (VO_2 max) – the maximum rate of oxygen the body can use during exercise. The study results showed an 11 per cent increase in VO_2 max compared to the placebo group and suggested that ongoing supplementation could increase benefits.

In 2010, another study tested the aerobic performance of twenty healthy adults aged between fifty and seventy-five years.[17] After twelve weeks of supplementing with extracts from a strain of *O. sinensis* called Cs-4, they showed a 10.5 per cent increase in their metabolic threshold. This indicated that they could exercise at a higher level without fatigue. There was also an 8.5 per cent increase in their ventilatory threshold, which meant a delay in the onset of lactic acid build-up that leads to muscle pain and cramps.

A 2016 study using *O. sinensis* showed benefits of increasing libido and sexual performance in both men and women.[18] Notably, when administered to twenty-two men for eight weeks, the data also showed a 33 per cent increase in sperm count, a 29 per cent decrease in sperm malformations, and a 79 per cent increase in the survival rate of sperm. These studies support the long held belief in China that *O. sinensis* can improve energy, vitality, endurance and sexual libido.

FUNGI AS A CANCER SUPPORTER

Our DNA encodes each cell in our body with instructions. The cells are told what to do, when to divide and when to die. The sheer number of cells constantly being replicated means that this process can go awry. Some cells stop following instructions, instead dividing rapidly and forming abnormal cells that can accumulate into a mass of tissue called a tumour. The good news is that, when the immune system is functioning properly, immune cells sweep up these damaged or malfunctioning cells and dispose of them. On the unfortunate occasion that these cells are not detected by the immune system and are allowed to proliferate, cancer starts.

Immunotherapy is the latest form of cancer therapy aimed at triggering the body's immune system to destroy cancer cells. Unlike chemotherapy and radiation therapy, which kill cancer cells directly, immunotherapy jump starts the body's powerful immune system to do the job it's designed for. Immunotherapy drugs harness the immune system's natural power to recognise, remember and eliminate cancer cells.

Immunotherapy is also used with other treatments to further improve results. Beta-glucans from medicinal mushrooms are known immune system potentiators. Increased activity in the immune cells increases the body's potential to respond to cancerous cells and helps prevent cell mutation. Human clinical trials using fungi for cancer support programs are promising, with beta-glucans proving to help the immune response during and after cancer treatments. These compounds found in medicinal mushrooms are called immunoceuticals. They have been used as conventional cancer treatments in Asia for the last thirty years.

Trametes versicolor, or turkey tail, is considered one of the most potent medicinal mushrooms. In the 1960s, a beta-glucan called polysaccharide krestin (PSK) was isolated from the mycelium of *T. versicolor* by Japanese scientists. Human clinical trials began in the 1970s, using PSK as a supporting medicine to create a stronger immune response for people undergoing surgery, radiation therapy and chemotherapy. The study found that 'PSK significantly extended survival at five years or beyond in cancers of the stomach, colon-rectum, esophagus, nasopharynx, and lung (non-small cell types), and a subset of breast cancers'.[19] In 1977, PSK was approved as a prescription drug by the Japanese Ministry of Medicine for this use. Ten years later, it accounted for one-quarter of anti-cancer drugs sales in Japan.

In the 1980s, Chinese scientists isolated a similar compound called polysaccharide peptide (PSP) from *T. versicolor*. In double-blind trials, PSP significantly extended five-year survival rates in esophageal cancer patients. It improved quality of life, provided substantial pain relief and enhanced immune status in 70 to 97 per cent of patients with cancers of the stomach, esophagus, lung, ovary and cervix.[20]

In addition to boosting immune cell production, PSK and PSP also help relieve the side effects of conventional cancer treatments, such as nausea and fatigue from chemotherapy. Both PSP and PSK are well-tolerated and patients reported little to no side effects.

While fungi are not used in first-line cancer treatment therapies, they are powerful allies when used as part of a holistic health program.

FUNGI AS A FACTORY

Scientists can also use fungi as microscopic factories to manufacture drugs that would otherwise have to be derived from nature. Just as yeast, a single-celled fungus, is used to brew alcohol, scientists can use yeast to brew complex medicines. In 2018, Stanford engineers genetically reprogrammed the cell of *Saccharomyces cerevisiae*, the hardworking brewer's yeast, to create noscapine, a cough suppressant typically produced from opium poppies.[21]

By instructing yeast to make essential drugs, production efficiency increases many times over and creates a more reliable way to source raw materials for medicine rather than relying on nature. The use of fungal factories also addresses the ecological and sustainability concerns of producing medicines directly from plants and fungi, which uses up these precious resources. Yeasts are now engineered to manufacture a range of life-saving drugs from insulin to vaccines, and this is just the beginning.

RIGHT
Typically found growing on hardwood trees, the sporing body of *Hericium erinaceus* is striking, with its icicle-like cascading white teeth.

3.4

CONSUMING MEDICINAL MUSHROOMS AND SUPPLEMENTS

Medicinal mushroom products have flooded the wellness and supplements market and are widely available online and in health food stores. Many of them use impressive marketing language on their packaging, but not all products are created equally. Cultivation methods, the part of the fungus that is used, growing mediums and extraction processes all affect the quality. Before using fungi as a dietary supplement, try incorporating mushrooms as a functional food first.

FOOD IS MEDICINE

Food is our best tool to prevent illnesses. The right diet bolsters our immune system and leaves us flush with vitamins and minerals – mushrooms, for example, contain an impressive amount of dietary fibre, antioxidants, vitamin D, B vitamins, protein and probiotics. However, humans cannot digest raw and undercooked mushrooms of most species. Medicinal compounds such as beta-glucans are bound up by chitin in the cell walls of fungi, but heating releases these active compounds and makes them more soluble as they pass through our digestive tract. This is why medicinal mushrooms are traditionally boiled for hours before consumption.

The good news is that increasingly accessible 'culinary' varieties, such as shiitake, enoki and oyster mushrooms, offer plenty of health benefits without the need for hours-long preparation – so include them into your diet by grilling them, sauteing them, or adding them to soups.

SUPPLEMENTS

For people who are looking for specific medicinal benefits, mushroom supplements (or nutraceuticals) offer a welcome alternative to conventional Western medicines. Made using compounds extracted from the sporing body or the mycelium, they boast low toxicity, even at high doses, and have very few, if any, side effects. But they are not panaceas. They are not a substitute for good quality sleep, a nutritious diet, regular exercise or other healthy lifestyle choices. Nor are they a quick fix; they are a long-term investment.

Since medicinal mushrooms work in broader and less symptom-focused ways, it takes time to notice the changes. Many herbal practitioners recommend taking medicinal mushrooms for at least three months to ensure the benefits take effect.

Herbal practitioners also often suggest switching between species or combining species so that different areas of your body are targeted. To date, there is no concluding study that suggests the immune system builds tolerance to the active compounds, so switching is not essential. For beginners, take your chosen mushroom supplement for a few weeks, then observe any effects.

SPORING BODY AND MYCELIUM

There is much debate about which part of the fungus is best used in supplements: the mycelium or the sporing body. However, this is not the right question to ask. A more important question is how much of the active compounds, such as beta-glucans, are bioavailable in a supplement. The form of the mushroom supplement is not important – the marker of quality is the bioavailability of the medicinal compounds.

Mycelium contains beneficial compounds that are no better or worse than those found in sporing bodies. Studies show that beta-glucans can be found in both the sporing body and mycelium, but the concentration can be higher in sporing bodies. The same research found that many mycelium supplements on the market are grown on brown rice grain,[22] which often ends up processed with the mycelium to create the final product, lowering the proportion that is actual fungi. Unfortunately, many products advertised as medicinal mushrooms are filled with starch from grains, which lowers their medicinal potency. When purchasing products, look at the percentage of beta-glucans on the label. Unfortunately, very few companies provide this.

Finally, it's important to note that when researching and buying mushroom supplements, it's common to see references to the 'fruiting body'. Fruiting body and sporing body are just different names for the same thing, both refer to the mushroom part of a fungus. While fruiting body is still widely used, particularly in the wellness and wellbeing industry, sporing body is generally considered the preferred term in mycology, hence our preference for it throughout this book.

HOW TO MAKE YOUR OWN SUPPLEMENTS

The most reliable, affordable and sustainable option when it comes to nutraceuticals is to grow your own mushrooms and extract your own medicines. Cultivating mushrooms at home using low-tech methods is the democratisation of health in action. Here are some tried and true techniques for accessing the active compounds from sporing bodies, starting with the easiest method and increasing in complexity:

Hot water extracts

Adding hot water to mushrooms is a simple way to extract medicinal compounds. Beta-glucans in mushrooms are soluble in water. Contrary to popular belief, hot water extraction will also draw out some triterpenes. Jeff Chilton explains: 'All you have to do is get yourself some actual reishi mushrooms, chop them up, throw them in hot water and boil them up for three hours. See how bitter it tastes? That's because most of the triterpenes are coming out.'[23]

1   Chop the mushrooms (fresh or dried) into small pieces and simmer in water for at least 60 minutes at a minimum of 70 °C. Use a 10:1 ratio of water to mushrooms for best results. The longer the simmering time, the more concentrated the extract will be.

2   Once the liquid turns a dark colour, strain the mixture to separate the liquid extract from the remaining mushrooms. The liquid extract is ready to be consumed.

3   This process can be repeated with new water until the mushrooms stop colouring the liquid.

4   Consume as a tea, in a soup, or any way you prefer. The recommended dosage is 1–2 ml drops taken once or twice a day, as preferred.

Tinctures

A tincture is an extract that is dissolved in alcohol to preserve the medicinal compounds in the mushrooms. Some terpene compounds in medicinal mushrooms are insoluble in water, so they are best extracted using alcohol. Alcohol absorbs more fat compounds (terpenes) than water and increases the shelf life while maintaining potency.

Tinctures are generally made using a cold extraction process – soaking the mushrooms in alcohol for two weeks and then straining. This method dates back 1000 years, when Egyptians created tinctures by soaking herbs in alcohol.

1 Chop the mushrooms (fresh or rehydrated) into small pieces and add them into a blender. Cover the mushrooms with a high-proof, food-grade alcohol (at least 80-proof, but 190-proof is preferred).

2 Blend this mixture for 20 seconds.

3 Pour the mixture into a jar. Add more alcohol to ensure the mushrooms are covered by at least 3 cm.

4 Close the jar tight and let it sit for two weeks. Shaking the jar daily helps the extraction process.

5 Strain the mixture and pour the liquid tincture into a dropper bottle.

6 Consume by placing drops under the tongue or mixing them in tea, coffee or other drinks. The recommended dosage is 1–2 ml drops taken once or twice a day, as preferred.

Double-extract tinctures

Combining water and alcohol extraction methods creates a more potent tincture. Most mushroom supplements on the market are double (or dual) extracted. This method is one of the best ways to receive greater benefits from medicinal mushrooms.

1 Follow the instructions for making a hot water extract.

2 Using the mushroom remnants left over from Step 1, follow the instructions for making a tincture.

3 Combine the two liquids into a dropper bottle.

4 Consume by placing drops under the tongue or mixing them in tea, coffee or other drinks. The recommended dosage is 1–2 ml drops taken once or twice a day, as preferred.

Concentrated extract powder

Making concentrated extract powders requires a few extra steps. The compounds need to be extracted in liquid first before they are ground for consumption. Powders are not made from ground-up dried sporing bodies – the medicinal compounds are not bioavailable in this form.

1 Cut the mushrooms into small pieces and simmer in a pot with water (use a 10:1 ratio of water to mushroom) for a few hours.

2 Once the mixture cools, blend it into a batter.

3 Heat the batter in a dehydrator or bake it lightly in an oven with the door open so the moisture can dissipate.

4 Grind the resulting dry pieces into a fine powder.

5 Consume by adding powder into tea, coffee or other drinks. The recommended dosage is 5 ml, approximately 4 grams, taken once a day.

Considerations when buying supplements

JEFF CHILTON

The market for medicinal mushroom supplements has experienced explosive growth, but there are few quality control standards in place. Supplements may be expensive, but they generally do not contain the same amounts of beneficial compounds as the substances used in human clinical studies. These starter points may help you evaluate mushroom supplements before purchase.

Sourcing

It is important to know whether the product you are purchasing is made from pure fungi. Most mushroom supplements are produced in China, as it produces 85 per cent of the world's mushrooms and has a long history using medicinal mushrooms. But not all mushrooms are grown to a high-quality standard, whether they're grown in Asia, the Americas or Europe, so some research is necessary.

Substrate and extraction medium

High-quality mushrooms are grown on a wood medium, which is important for producing the desired medicinal compounds. Mycelium grown in liquid and harvested can provide similar medicinal benefits. But you must be careful about products that grow mycelium on grains. Grains contain lots of starch, and once a grain is colonised it can be very difficult to separate the mycelium from the substrate, so the two are often processed together. This means that large quantities of starch can end up as filler in your mushroom supplements, reducing the medicinal value of the product.

Many US companies are producing mycelium on sterile grains and calling their products 'mushroom'. Look very carefully at the label to see if it says myceliated rice or oats – this is mycelium grown on grains, which is not what you want. For best results, look for pure mycelium or mushroom extracts. You can find out if your products contain grains with the iodine test described below.

Extraction method

If you can, take the time to find out how the nutrients were extracted for each product. Most mushroom supplements are hot water extracts, as this is sufficient to concentrate the important beta-glucans. This is known as single extraction. But mushrooms high in triterpenes, such as reishi (*Ganoderma lingzhi*) and chaga (*Inonotus obliquus*) should be dual extracts – combining hot water extraction and alcohol – because some of the valuable triterpenes are insoluble in water.

Jeff Chilton studied ethnomycology at the University of Washington in the late 1960s, and in 1973 he began a ten-year career as a large-scale commercial mushroom grower. Jeff is the co-author of *The Mushroom Cultivator*, published in 1983. In 1989, he established Nammex, the first company to supply medicinal mushroom extracts to the nutritional supplements industry. In 1997, he organised the first organic certification workshop for mushroom production in China. Jeff is also a founding member of the World Society for Mushroom Biology and Mushroom Products and is a Member of the International Society for Mushroom Science.

Watch out for simple unprocessed dried mushroom powder, as it is low in beneficial compounds. If you are purchasing a tincture, you should check how much water or alcohol makes up the bottle's contents. Tinctures are mostly liquid and can be very diluted compared to an extract powder.

Potency of active compound

All mushrooms have polysaccharides in them, and the important ones are called beta-glucans. Look for the actual amounts of beta-glucans on the product labels. Beware: a high percentage of polysaccharides on a product label isn't correlated with a high percentage of beta-glucans. Products are often mixed with polysaccharide carriers, which can be used as stabilisers, but these are alpha-glucans and starch. Polysaccharides are not a valid quality assurance measure. All high-quality products should clearly state the percentage of beta-glucans – 20 per cent is a good benchmark.

Triterpenes are only found in select species, such as reishi (*G. lingzhi*) and chaga (*I. obliquus*). Some mushroom supplements disclose the levels of triterpenes – the higher the level, the better the quality. Tests conducted on commercial reishi products showed triterpene levels ranging from undetectable to 7.8 per cent. Triterpenes are what make reishi bitter – if you can't taste concentrated bitterness, it's not reishi.

Testing using the Megazyme and iodine test

The US Food and Drug Administration (FDA) regulates the clean, safe manufacturing of products, but rarely intervenes regarding efficacy. Quality control is still in its early days. Some companies test their products using the internationally recognised Megazyme test, which tests for levels of beta-glucans and starch. Before purchasing any product, find out if test reports are available.

If you want to take testing into your own hands, an iodine test is an easy way to confirm if your powder contains starch. Stir in 2 to 3 grams or six capsules of the powder into ¼ cup of water and mix it well for a minute or two. Stir in ten drops of a standard iodine solution. A pure mushroom supplement should only change to the colour of the iodine. If the liquid turns black or blue, this signifies a high starch content.

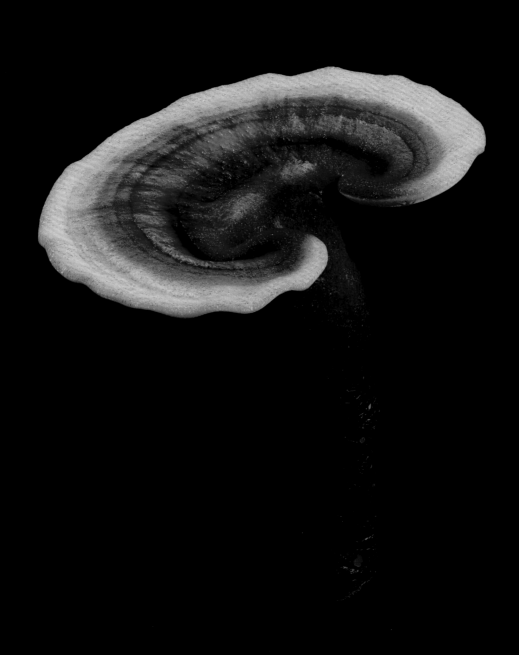

GANODERMA LINGZHI

COMMON NAMES

Ling zhi ('divine mushroom' in Chinese),
reishi ('spirit mushroom' in Japanese),
mannen-take ('10,000-year-old mushroom'
in Japanese)

FAMILY	Ganodermataceae
GENUS	Ganoderma
SPECIES EPITHET	*lingzhi*

Ganoderma means 'shiny skin', a reference to its lustrous red surface that fans open and grows in clusters or rows. Like all polypores, it lacks gills on the underside, and instead releases its spores via fine pores. *Ganoderma lingzhi* is rare and hard to find, so wild harvests are sold for more than US$500 per kilogram. However, thanks to successful cultivation on hardwood logs and sawdust, it is accessible to anyone looking to strengthen their immune system without breaking the bank.

HISTORY AND CULTURE

G. lingzhi has inspired oriental art, medicine, spirituality and myths for at least 2500 years. It was called the 'mushroom of immortality', as it was believed that it could strengthen the heart, calm the spirit and replenish the energy of the body, mind and spirit. *G. lingzhi* was identified as *G. lucidum* by Western mycologists until DNA analysis showed that the latter is a European species that doesn't grow in Asia. The difference is slight – the core layer beneath *G. lingzhi* is yellow while *G. lucidum* is white. Both are equally medicinal.

PROPERTIES

EDIBLE

Yes, but very tough, woody and bitter so not recommended in raw form. Usually ground and boiled into a tea or tincture, rather than cooked as food.

MEDICINAL

Yes. Contains over 300 medicinal compounds and is particularly high in beta-glucans. Has a reputation as a cure-all in Asia due to its extensive historical use. Being researched in the development of drugs for areas including antibiotics, antivirals, anti-cancer compounds, blood pressure medication and antioxidants.

PSYCHOACTIVE

No.

ENVIRONMENTAL REMEDIATION

No, but its European cousin, *G. lucidum*, has been used to remediate heavy metals, insecticides and petroleum hydrocarbons with success.[24]

SPORING BODY CHARACTERISTICS

CAP

2–30 cm wide
4–8 cm thick
Circular to fan-shaped
Bands of brown, red, orange, yellow and white
Surface grooved, varnished
Texture hard or leathery

PORES

White to brown
4–7 per mm

STIPE

3–15 cm tall
0.5–4 cm thick
Dark red to red-black
May be varnished
May be absent

SPORES

Red-brown
Oval

FIELD DESCRIPTION

HABITAT

Grows on decaying deciduous trees, particularly maple.

DISTRIBUTION RANGE

Asia.

SEASON

All year.

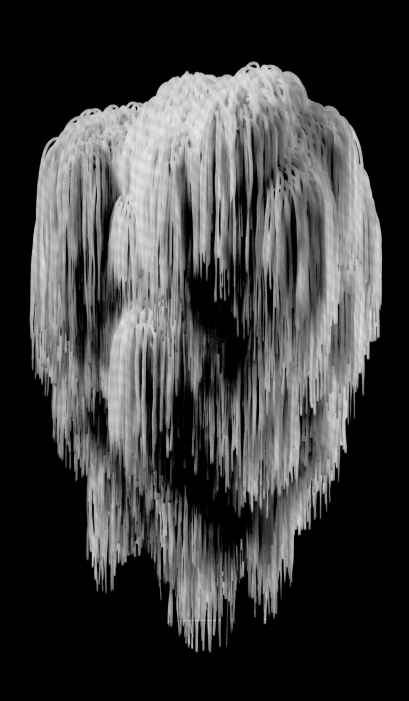

HERICIUM ERINACEUS

COMMON NAMES

Lion's mane, old man's beard, monkey's head, bearded tooth, bearded hedgehog, pom pom, *hou tou gu* ('monkey head mushroom' in Chinese), *yamabushiitake* ('bearded tooth fungus' in Japanese)

FAMILY	Hericiaceae
GENUS	Hericium
SPECIES EPITHET	erinaceus

In the wild, *Hericium erinaceus* primarily grows on decaying trees, but it can also be parasitic on living trees. The white pom poms with dense, dangling icicles are hard to miss. It is not only beautiful to look at but also delicious, and has a rich history of medicinal use. All species in the *Hericium* genus are edible and best when young, tender and pure white.

HISTORY AND CULTURE

H. erinaceus has been used traditionally for hundreds of years in China and Japan as a general health tonic. Advances in cultivation methods in the last two decades have made it more available for research and consumption. Today, it can be found fresh from farmers markets, dried from Asian grocers and in powdered form from health stores – but make sure it is pure mycelium or sporing body extracts if you are after its medicinal effects.

PROPERTIES

EDIBLE

Yes, very. Has a sweet, nutty taste and the texture of seafood, such as lobster or crab.

NUTRITIONAL PROFILE

A raw 100-gram serving contains 35 calories, composed of 89 % water, 8 % carbohydrate, 2 % protein and less than 1 % fat. Rich in vitamins, providing 20 % RDI of B vitamins, and minerals such as iron and potassium.

PSYCHOACTIVE

No.

MEDICINAL

Yes. Contains powerful terpenes and beta-glucans. It has been tested in human clinical trials for neuroprotective and neuroregenerative properties. There's no conclusive evidence yet, but it's widely sold as a supplement for neurological health support and immune system stimulantion.

ENVIRONMENTAL REMEDIATION

No.

SPORING BODY CHARACTERISTICS

BODY

10–75 cm wide and tall
Round
White to yellow

TEETH

Soft hanging spines
White to yellow
1–6 cm

SPORES

White
Round

FIELD DESCRIPTION

HABITAT

Grows on open wounds of hardwood trees, such as oak, beech, walnut and maple.

DISTRIBUTION RANGE

Found throughout the northern hemisphere. Native to North America, Europe and Asia.

SEASON

Summer and autumn.

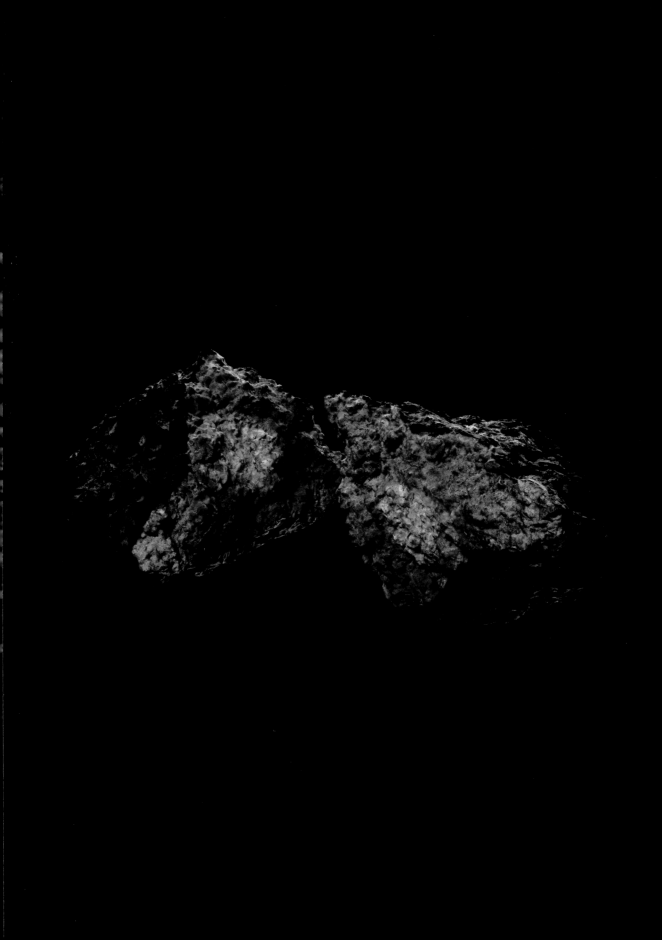

INONOTUS OBLIQUUS

COMMON NAMES

Chaga, clinker polypore, birch clinker, false tinder conk, woodpecker tea, *kabanoanatake* ('birch mushroom' in Japanese)

FAMILY	*Hymenochaetaceae*
GENUS	*Inonotus*
SPECIES EPITHET	*obliquus*

Inonotus obliquus grows as an unassuming black mass on the sides of trees. This black mass is not a sporing body. It grows in the form of a sterile conk to survive the harsh climates of its native regions. The conk is a mix of fungal sclerotia, which is compacted mycelium, and tree tissue. When cracked open, it reveals a beautiful golden texture. It is also medicinal gold – it has been used in tonics for hundreds of years by the local communities where it grows.

HISTORY AND CULTURE

I. obliquus is not just a medicinal powerhouse, it also tastes like chocolate and coffee. It has a strong, earthy, slightly bitter, nutty, vanilla flavour, which was leveraged in World War II during coffee and sugar shortages. *I. obliquus* was ground and steeped in water to make a drink that was substituted for coffee due to its energising properties. Today it is a popular health supplement worldwide, used for its immune support and antioxidant functions. For those sensitive to caffeine, it remains a great alternative to coffee – purchase chunks from a reputable supplier and brew your own.

PROPERTIES

EDIBLE

Not in raw form. Traditionally ground and boiled into a tea or tincture.

MEDICINAL

Yes. Used traditionally in Russia and northern Europe as a general health tonic and a treatment for cancers, liver and heart disease and digestive issues. In Russia, it has also been developed into a licensed medication called Befungin.

PSYCHOACTIVE

No.

ENVIRONMENTAL REMEDIATION

No.

SPORING BODY CHARACTERISTICS

BODY

25–40 cm wide and tall
Irregularly shaped
Black
Surface becoming cracked
Texture hard

PORES

Dark brown to white
3–5 per mm

SPORES

White
Oval

FIELD DESCRIPTION

HABITAT

A parasitic fungus that attacks the bark of older trees. Grows almost exclusively on birch trees, but has also been found on elm and hornbeam trees.

DISTRIBUTION RANGE

Found in cold climate regions towards the Arctic, such as Russia, eastern and central Europe, Canada and north-eastern America.

SEASON

Found throughout the year.

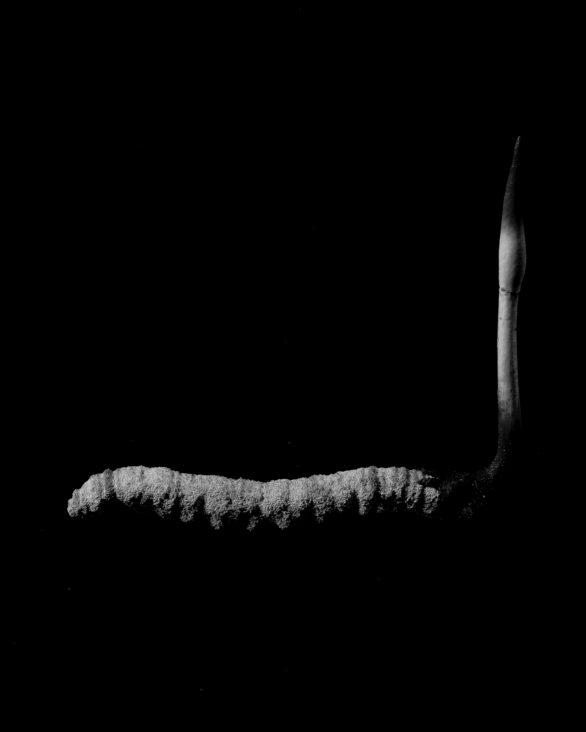

OPHIOCORDYCEPS SINENSIS

COMMON NAMES

Caterpillar fungus, *dong chong xia cao* ('winter worm, summer grass' in Chinese)

FAMILY	Ophiocordycipitaceae
GENUS	Ophiocordyceps
SPECIES EPITHET	*sinensis*

Ophiocordyceps is a gory genus. It evolved to feed on live insects, consuming their internal organs then taking over their bodies. *Ophiocordyceps sinensis* has a particular taste for ghost moth caterpillars. In China, it is called 'winter worm, summer grass' because spores germinate in winter and mycelium grows throughout the caterpillar's body. When summer comes, a club-shaped sporing body bursts out of the caterpillar's head and protrudes from the ground so that foragers can locate it. *O. sinensis* is then consumed whole, caterpillar body and all.

HISTORY AND CULTURE

O. sinensis has a rich medicinal history in Asia. Growing demand and shrinking supply makes it the world's most expensive parasite – it is worth more than gold. Families in and near the Himalayan mountain ranges uproot their lives in May and June to forage for it. They crawl on their elbows and knees, inch by inch across the plateau, to locate the club-shaped sporing body. Harvesting *O. sinensis* is an important livelihood for hundreds of thousands of people. Two months of work can pay the family's living expenses for the rest of the year. Unfortunately, excessive harvesting and a warming climate are affecting the delicate ecosystem, and yet there is no conservation policy in place.

PROPERTIES

EDIBLE

Yes, although it's used for medicinal purposes rather than gastronomy.

MEDICINAL

Yes. Used extensively in Asia to improve energy, vitality and athletic performance. Also called Himalayan viagra, it is used as an aphrodisiac.

PSYCHOACTIVE

No.

ENVIRONMENTAL REMEDIATION

No.

SPORING BODY CHARACTERISTICS

HEAD

1.5–2.5 cm long
3–5 mm wide
Cylindrical to club-shaped, pointed
Yellow to brown
Dry, granular from openings of tiny sacs

STIPE

2.5–8.5 cm long
0.15–0.3 cm thick
Yellow, brown or black
Smooth, ridged
Attached to mummified caterpillar

SPORES

White
Oval

FIELD DESCRIPTION

HABITAT

In soils at high altitudes. Spores infect caterpillars when they are feeding on plant roots in soil.

DISTRIBUTION RANGE

Found at altitudes above 3500 metres in the Himalayan mountain ranges.

SEASON

Summer.

TRAMETES VERSICOLOR

COMMON NAMES

Turkey tail, *yun zhi* ('cloud mushroom' in Chinese), *kawaratake* ('tile mushroom' in Japanese)

FAMILY	*Polyporaceae*
GENUS	*Trametes*
SPECIES EPITHET	*versicolor*

Versicolor means 'having many colours'. True to its name, *Trametes versicolor* fans open with varying bands of brown, tan, blue, grey or white, which vividly resemble a turkey's tail. It grows on wood in the shape of shelves or brackets, usually in many clusters or rows. It is a polypore, which means that it releases spores from pores on its underside instead of from gills. Look down to find them – they are saprophytes and prefer decomposing logs.

HISTORY AND CULTURE

The healing properties of *T. versicolor* were documented as early as the 16th century in the *Compendium of Materia Medica* during the Ming dynasty in China. It remains one of the most respected natural medicines, used to boost the immune system, support detoxification and improve energy levels of both the physical and spiritual body. Although it has been a clinical treatment for cancers in Japan and China since the 1970s, results from human clinical trials remain inconclusive in the Western world.

PROPERTIES

EDIBLE

Yes. The leathery flesh can be chewed raw like gum. However, the texture is thick and tough, so it is best consumed in teas or soups after boiling in hot water.

MEDICINAL

Yes. It is the most clinically tested medicinal mushroom. It contains polysaccharides that are used as anti-cancer drugs in Japan and China, and high levels of beta-glucans. Available in health food stores in powdered form to help modulate the immune system, support liver health and improve gut health.

PSYCHOACTIVE

No.

ENVIRONMENTAL REMEDIATION

Yes, it is a white rot fungus. Produces extremely strong enzymes that can break down pollutants such as heavy metals, pesticides, pharmaceuticals and hydrocarbons. Tested extensively and shows great promise in mycoremediation.[25]

SPORING BODY CHARACTERISTICS

CAP

2–10 cm wide
Fan-shaped
Bands of brown, tan, blue, grey and white
Surface velvety
Texture hard or leathery

PORES

White to yellow
3–5 per mm

SPORES

White to yellow
Cylindrical

FIELD DESCRIPTION

HABITAT

Grows on dead or decaying logs of hardwood trees, such as oak, beech, maple and birch. Occasionally found on conifers.

DISTRIBUTION RANGE

Widespread and very common in forests on every continent except Antarctica.

SEASON

All year.

PSYCHEDELICS

Fungi allow us to form new connections in our brain and dissolve the rigid boundaries of our minds. A study using the fungal compound psilocybin found that 67 per cent of the participants ranked their experience as one of the top five most meaningful things they have ever done, on par with the birth of a child, death of a loved one, or marriage.[1]

FUNGI CAN FREE US

There are profound questions about our existence that govern our lives, whether we ask them or not. Around 14 billion years ago, stars turned supernova and showered the universe with the building blocks from which we are made.[2] We not only triumphed over improbable odds to be alive, but we also received the spark of consciousness. We are 97 per cent stardust — and became aware of our existence.[3]

Only through consciousness does our world come alive, as a series of fleeting images changing moment by moment. Our brain, with its billions of neurons, fills in the gaps between those snapshots. What we call everyday reality is a projection of stories in the theatre of our minds. We live with world views that we are unaware of, hypnotised by the assembling stories that have been handed to us. So who are we, beneath the layers of cultural and social conditioning? What is our purpose? Anyone interested in unravelling the complexity that is the human condition should be interested in psychedelics. They provide a doorway into the exploration of consciousness and our shared experience on Earth.

As conscious beings, we can think and strategise; we are masters of our physical world. But our inner worlds are crumbling, with epidemic levels of loneliness, depression and anxiety. Business as usual is simply not enough. The prevailing rational and anti-spiritual attitude of our time does not satisfy our existential longing for meaning. Just as we are shaped by our experiences, we are also shaped by things we never contemplate.

Consider for a moment that, beneath the urgency of everyday life, there is boundless love, peace and connection dwelling in the depths of your being. Psychedelics offer us a glimpse into that possibility. These states of enlightened ecstasy were once only available to a fortunate few, but now they are accessible through a psilocybin mushroom or LSD experience. Psychedelics allow us to transcend our individual identities, and free us from all the ways our minds create stories that do not serve us. They dissolve boundaries within us and between us. In this space, we can rewrite entire value systems and beliefs. And, contrary to popular belief, psychedelics are safe when they are used properly. They can heal trauma with just one experience, are naturally occurring, non-toxic and non-addictive. What other compounds can do this?

Classic psychedelic compounds include mescaline, DMT, psilocybin and LSD. The last two come from the fungi kingdom. Before the hippie revolution of the 1960s, before modern science and even before the written word, our ancestors used psychedelics in sacred ceremonies. At the core of these traditions, psychedelics were used to reconnect with the divine. It was a way to reveal the god, or the soul, within.

Dr Humphry Osmond, an English psychiatrist known for his pioneering research into LSD, coined the word 'psychedelic' in his rhyme 'to fathom Hell or soar angelic, just take a pinch of psychedelic'.[4] *Psyche* means 'soul' in Greek, and *delos* means 'manifesting'; 'psychedelic' therefore means 'soul manifesting' or 'to make the soul appear'. More recently, the term 'entheogen', meaning 'revealing the god within', has been used interchangeably with 'psychedelic' in ceremonial or religious contexts.

Science has shown that, at high doses, psychedelics can reliably cause spiritual experiences. However, concepts such as spirit and soul can be abstract for people brought up in Western societies, because they can't be seen, touched, bought or rationalised. As such, the word 'psyche' has become synonymous with 'mind', and psychedelics are often defined as 'mind manifesting'.

It's time for our society to reconnect with the spiritual dimension of our existence. As Carl Sagan, the celebrated American astronomer, Pulitzer Prize–winning author and populariser of science, reasoned, 'Science is not only compatible with spirituality; it is a profound source of spirituality. When we recognise our place in an immensity of light-years and in the passage of ages, when we grasp the intricacy, beauty, and subtlety of life, then that soaring feeling, that sense of elation and humility combined, is surely spiritual. The notion that science and spirituality are somehow mutually exclusive does a disservice to both.'[5]

In modern history, psychedelics became intertwined with social movements and political motivations and were met with worldwide prohibition by the 1970s. We have lost decades of opportunity to develop a culture that guides its citizens to use psychedelics in productive and useful ways. Thankfully, the light of psychedelics is flickering brightly once again. Dedicated researchers, doctors and advocates are paving the road to widespread acceptance by using legitimate scientific studies. Psilocybin has reached phase II clinical trials in the US, and more psychedelic medicine companies are going public. Decriminalisation and legalisation movements are making great strides, and the psychedelic renaissance is set for a massive boom.

This curiosity about the effects of psychedelics is also filtering through to the mainstream. Everyday people are turning to them to make sense of their existence – to find out who they are, who they are not and how they can recalibrate their lives.

Indigenous cultures preserved the knowledge of psychedelics through generations, and we are privileged to receive their wisdom. We need to approach psychedelics with respect. They are a sacred tool for accessing repressed emotions, thoughts and memories, and exploring altered states of consciousness. We have been endowed with the ability to experience and make sense of the world. The adventures of inner space await you.

Psychedelics allow us to transcend our individual identities, and free us from all the ways our minds create stories that do not serve us.

4.1

THE HISTORY OF FUNGI AS A PSYCHEDELIC

Shamanistic cultures have long revered Earth as a spiritual, living entity. They believe that everything – from rocks and rivers to plants and fungi – is alive and contains a spiritual essence. Sacred ceremonies are held to connect with spiritual planes of existence. Some tribes use fasting, drumming, dancing and chanting to induce ecstatic states. Some also use psychoactive fungi and plants in rituals led by the shaman or medicine man or woman. These ceremonies seek out the intrinsic desire to connect with states larger than ourselves, to reconnect with ancestors and gods. The shaman enters the spiritual realm and brings back knowledge to heal illnesses, manage interpersonal relationships and ensure the long-term wellbeing of the community.[6] This exploration of mystical states of consciousness has facilitated our growth as a species for a very long time.

PREHISTORY

While it is impossible to trace the origins of our partnership with psychoactive fungi, the Stoned Ape theory is a famous and enduring hypothesis put forward in 1992 by Terence McKenna, ethnobotanist and crusader for the democratisation of entheogens. Historians accept that our ancestors left the forest canopies to explore Africa 2 million years ago. McKenna proposes that we have been in biological symbiosis with psychoactive mushrooms since we left the trees for the harsh savanna. There, we encountered mind-expanding *Psilocybe* mushrooms in cattle dung and consumed them as a nutritious source of food. The evolutionary consequence of this dietary choice over millions of years gave rise to human cognition, self-awareness and creativity.

This could provide an explanation for the rapid increase in brain size that is unique to our species. While fascinating to contemplate, the earliest archaeological evidence of the human–mushroom partnership dates back to 10,000 BCE and comes from cave paintings in the Sahara Desert. Palaeolithic hunter-gatherers used art to convey their stories, beliefs and rituals. One cave painting depicts a looming figure with mushrooms growing throughout the outline of its body, likely to represent the powerful influence of mushrooms when consumed. Mushroom references are shrouded in mystery, but they are embedded throughout ancient cultures and religions.

ANCIENT HISTORY

The Eleusinian Mysteries in ancient Greece were sacred ceremonies held regularly near Athens. They were started around 1500 BCE by the well-respected cult of Demeter, goddess of agriculture, fertility and marriage.[7] These were the holiest of all celebrations in Greece, attended by men and women from all ranks – priests, noble families and slaves. Upon arriving at the Temple of Demeter, participants in the religious ritual drank a potion called *kykeon*.

Greek philosopher Plato described his visions at the Eleusinian Mysteries in his dialogues. In *Phaedrus*, he wrote an ancient trip report: 'we stood as initiates in a pure light, and we were pure and not entombed in this so-called body that we carry around with us, imprisoned like an oyster in its shell.'[8] This was the basis of his concept of dualism: that the physical body and the soul are separate entities, and that the soul lives on after death.

The contents of *kykeon* fascinate historians, who agree that there was a psychedelic component that facilitated the experience of the divine. The particular psychedelic used is still fiercely debated. The prevailing theory is that *kykeon* included a compound derived from ergot, a fungal mass that forms when the parasitic fungus *Claviceps purpurea* colonises grain. Discoveries of ergot remnants in a temple dedicated to two Eleusinian goddesses, excavated in Spain, provide legitimacy for the theory.[9] The Eleusinian Mysteries ended in 392 CE, when Rome ruled Greece and imposed Christianity as a state religion.[10]

Indigenous civilisations of Central and South America, such as the Maya, Aztec, Olmec and Zapotec, used psychoactive fungi and plants in healing and religious ceremonies for thousands of years.[11] These entheogens were sacred because of the states of ecstatic awe that they could induce. Mushrooms, in particular, were worshipped and called *teonanácatl* in the Nahuatl language of the Aztecs, meaning 'flesh of the gods'. The Aztecs held night-long ceremonies using *Psilocybe mexicana*, or other *Psilocybe* species, which were available after rains most times of the year.

Change came to the Aztecs in the 16th century when the Spanish conquistadors, led by Hernán Cortés, descended upon them. The conquering Spanish army denounced the Aztec's ritualistic use of fungi and plants as it threatened their Catholic beliefs. The conquistadors argued that the Indigenous practice of using psychoactive substances to achieve godhood was demonic and blasphemous. One cleric wrote back to the king, 'When they are eaten or drunk, they intoxicate, depriving those who partake of them of their senses and making them believe a thousand absurdities.'[12]

This last native Mesoamerican empire fell to the Spaniards in 1521. Their school systems and cultural traditions, including the use of *Psilocybe* mushrooms, were suppressed and replaced with Catholicism. This colonisation severed the Aztecs from their narratives, land and communities. But the wisdom of the mushrooms endured by going underground. For hundreds of years, these divine mushrooms remained the stuff of myths and legends. People continued to use them in secret, passing the knowledge from one generation to another.

BELOW

Psilocybe cubensis growing from their nutritious universe – a dung pile.

MODERN HISTORY

In the 19th century, Western researchers became interested in the traditions of Indigenous cultures. Dr Blasius Pablo Reko, an Austrian ethnobotanist, inspired the first wave of interest in psychoactive mushrooms. He was a scholar of the Indigenous use of medical plants and spent his career living with Indigenous groups of Mexico and learning from them. In 1936, he was the first to successfully identify *teonanácatl* as the psilocybin mushroom. Reko's groundbreaking work caught the attention of a young Harvard student, Richard Evans Schultes. Schultes soon joined Reko in his field research of native Mazatec culture in the state of Oaxaca in Mexico. In 1939, Schultes published his doctoral thesis 'The identification of teonanácatl, a narcotic basidiomycete of the Aztecs', shining a light on the hidden world of sacred fungi.[13]

One year earlier, on the other side of the Atlantic, a Swiss chemist called Albert Hoffman had accidentally synthesised LSD. He was searching for new medicines in the laboratories of Sandoz in Basel, Switzerland. Hoffman was working with ergot at the time. He combined its active compound, lysergic acid, with other organic molecules to see what would happen. In the twenty-fifth combination, he added diethylamide and synthesised lysergic acid diethylamide, abbreviated to LSD-25 for ease of reference in lab testing. It didn't show any medically valuable properties, so he moved on.

Whether through fate, intuition or 'a peculiar presentiment',[14] as Hoffman called it in his book *LSD, My Problem Child*, he felt compelled to investigate LSD-25 again five years later. On 19 April 1943, he self-administered what he thought was a tiny amount of LSD – 250 µg (double the now-standard dose). In his journal, he wrote: 'Beginning dizziness, feeling of anxiety, visual distortions, symptoms of ataxia, desire to laugh.'[15] Later, his lab assistant escorted him home on their bicycles due to restrictions on the use of cars during World War II. To this day, 19 April is affectionately celebrated as Bicycle Day to honour the anniversary of the world's first intentional LSD trip.

Recognising its profound and introspective nature, Hoffman foresaw LSD as a powerful tool for brain research and psychiatry. Sandoz began producing and distributing LSD and played a huge role in promoting the scientific development of the drug. Throughout the 1950s, it attracted academic attention across Europe, Canada and the US. LSD revolutionised the study of the brain and previously untreatable conditions, generating hundreds of scientific papers.[16]

1950s

The Mazatec village of Huautla de Jiménez is clustered in the mountains of northern Oaxaca. It overlooks a valley dominated by lush, steamy jungles. Despite their remoteness, Mazatec people had received visitors curious about their ancient traditions involving sacred mushrooms. Dr Valentina Wasson and her banker husband, Gordon Wasson, had a deep interest in the role of mushrooms in ancient

BELOW

Parasitic fungi from the *Claviceps* genus infect rye and form ergot, the purple-black mass.

cultures. Inspired by the work of Schultes, the Wassons made repeated visits to Mexico in the early 1950s in search of the sacred mushrooms.

In 1955, Gordon Wasson and his photographer, Allan Richardson, became the first Westerners to take part in the fabled Indigenous mushroom ceremony called *veladas*. It was led by María Sabina, the *curandera* (medicine woman) of Huautla de Jiménez. She used *Psilocybe* mushrooms to communicate with the divine in her night-time healing ceremonies. Wasson reflected after the ceremony, 'For the first time the word ecstasy took on real meaning. For the first time it did not mean someone else's state of mind.'[17] María Sabina shared her ancestors' sacred rituals on the condition that Wasson would not publicise her name or location.

Two years later, Gordon Wasson published an article titled 'Seeking the Magic Mushroom' in *LIFE* magazine.[18] Wasson's choice to use *LIFE*, the then-leading magazine in America, thrust 'magic mushrooms' into mainstream awareness. The article not only detailed the sacred ritual in a twenty-page spread but was accompanied by coloured photographs of María Sabina and the ceremony. For better or worse, this was a turning point for Huautla de Jiménez. The town was flooded with seekers of all kinds: artists, musicians, philosophers, hippies and scientists in search of *veladas* and the thrill of experiencing the divine. This forever altered the fabric of the Mazatec community, and María Sabina was ostracised for debasing the sanctity of their traditions. Today, however, she is considered a saint in her village and internationally, and 'a symbol of wisdom and love' for sharing her ancient Indigenous knowledge freely.[19]

In 1958, Albert Hoffman received samples of the sacred mushrooms from Wasson. He was the first to isolate the active compounds in *Psilocybe mexicana* and he named them psilocybin and psilocin. Hoffman created a synthetic version of psilocybin as pills. These were presented to María Sabina, who ingested them and confirmed that 'the spirit of the mushroom is in the pill'. With that, the effects of these ancient rituals were scientifically attributed to one compound: psilocybin. Soon after, Sandoz distributed psilocybin in pink 2 mg pills under the brand name Indocybin.

Psilocybin and LSD also piqued the interest of the US Central Intelligence Agency (CIA). Against the backdrop of the Cold War, they were searching for weapons for mind control and chemical warfare. For two decades, the CIA's MK-Ultra project tested psychedelics legally – and illegally – on citizens to understand its potential as a truth serum, interrogation tool and instrument for behavioural manipulation.[20]

1960s

Psychedelics also enthralled Timothy Leary, a prominent psychologist at Harvard University who would become the High Priest of LSD. After his first experience with psilocybin mushrooms in 1960, he later declared, 'I learned more in the six or seven hours of this experience than I had learned in all my years as a psychologist'. He said that, 'It was above all and without question the deepest religious experience of my life. I discovered that beauty, revelation, sensuality, the cellular history

of the past, God, the devil, all lie inside my body, outside my mind.'
In contrast, the ordinary state of mind was 'a static repetitive circuit'.[21]

It's understandable that Leary's psychedelic experience eclipsed
all his previous academic training and personal beliefs. Psychedelics
impart a noetic quality: a profound revelation that a significant truth –
the absolute truth – has been revealed. He had peered beyond the veil
of the socially constructed reality.

When Leary returned to Harvard after that summer, he created
the Harvard Psilocybin Project and ordered psilocybin from Sandoz.
Along with Richard Alpert, who would later be known as Ram Dass,
and graduate students Ralph Metzner, Gunther Weil and George
Litwin, the team set off to understand the frontiers of consciousness.
Ram Dass reflects on this time in his book *Be Here Now*: 'We were
exploring this inner realm of consciousness that we had been
theorising about all these years and suddenly we were travelling
in and through and around it.'[22]

Leary attracted artists, philosophers and scientists and shared
psychedelic experiences with notable figures such as Aldous Huxley,
Alan Watts, William S Burroughs and Carl Sagan. Through hundreds
of sessions, the trailblazing team coined the term 'set and setting',
which are huge determinants of the experience. 'Set' refers to the
mindset of the person taking the psychedelic. Having clear intentions,
being in a good headspace and respecting the sacred medicine can
significantly impact the experience. 'Setting' refers to the physical and
social environment. For an introspective journey, a safe environment
with trusted companions or guides creates a container for healing.

With these insights, the Harvard Psilocybin Project set up the
Concord Prison Experiment. The team wanted to see if psilocybin
could transform the prisoners' awareness to help them stay out of
prison, reducing the recidivism rate. They also conducted the famous
Good Friday Experiment, designed by Walter Pahnke, a graduate
student at Harvard Divinity School. The goal of the Good Friday
Experiment was to understand if psilocybin could stimulate the same
mystical experiences that occur naturally in religious states.[23]

These experiments were controversial. Colleagues questioned
the researchers' study methods, data analysis and tendency to take
psychedelics with study volunteers. Frequent psychedelic sessions,
which by this time included LSD, were also held outside the campus,
drawing criticism from other faculty members. They were accused
of blurring the line between researching psychedelics and promoting
their recreational use. The criticism from his colleagues didn't land well
with Leary's anti-authoritarian and outspoken disposition. Leary, with
the support of the prominent poet and writer Allen Ginsberg, wanted
to democratise access to psychedelics. They wanted regular people
outside the circles of privilege and academia to access greater states
of consciousness. The Harvard Psilocybin Project ended by 1963 when
both Leary and Alpert were more or less fired from Harvard.

They continued to push the psychedelic mindset to release human
potential from within the culture. Leary's urging to 'Turn on, tune in,
drop out' wasn't a message for youth to leave their schools, jobs and
families and do drugs, as the media had portrayed it. In its essence,
Leary meant: 'Turn on' – find a ritual which takes you out of your

thinking mind and into your true self. This does not have to be through psychedelics. 'Tune in' – harness and embody what you learn. Learn to express the new, truer version of yourself. 'Drop out' – choose to detach from the unconscious games of social convention. Become self-reliant by finding the freedom within.

Leary guided the counterculture through psychedelic experiences by co-authoring *The Psychedelic Experience* with Ralph Metzner and Richard Alpert, which was based on Eastern mysticism traditions, and writing *Psychedelic Prayers: And Other Meditations*, based on the classic Chinese text *Tao Te Ching*. Leary's message wasn't 'Do drugs'; it was 'Think for yourself and question authority'. With protest in the air and the US entering the Vietnam War, this was not a welcome ideology.

Even for those born after 1967's Summer of Love, the swinging, heady days of the 1960s echo in our collective memories. Flower children wore their hair long, donned thrifted prints and bell bottoms, listened to rock by the Grateful Dead and The Beatles and read novels from the Beat Generation. As a direct rebellion to the consumerist mindset of the 1950s, they celebrated their connection to the Earth rather than placing value on money, material possessions and status symbols. The counterculture was trying to answer a classic question: Is there an answer to the suffering in the world?

In a time when trust in the government reached a new low, disillusioned youth were looking for an antidote to the turbulent times they lived in. Psilocybin and LSD provided immediate answers: they shifted world views, turned on ecological awareness and cultivated a tribal intimacy. Psychedelics gave them a firsthand exploration of the big questions. They had a new lens to use to look at the nature of reality, life's meaning and how to live with peace and harmony. It was as if this renewed expression of ancient rituals returned a piece of the sacred to the youth. For centuries, wise Indigenous cultures have used sacred mushrooms for awakenings. These rites of passage are carefully guided, and the profound experiences they generated become deeply integrated into their way of life. These traditions are missing in Western society.

The goals of the 1960s influenced and continue to inspire decades of alternative lifestyles and values. The most significant social movements of our time – civil rights, feminism, gay rights, animal rights and the ecology movement – were all borne out of, or bolstered by, this countercultural mindset. This was cultivated in no small part by psychedelics.

Art and music festivals, such as Woodstock, were a melting pot of alternative lifestyles and ideas, and were decades ahead of their time. They advocated for yoga, meditation, complementary medicine, sustainable energy and wholesome foods. The influence on creativity and popular culture of the psychedelic explosion cannot be understated either. A kaleidoscope of novels, music, films and art was forged in this state of consciousness. Youthful idealism believed that, with time, love and peace would win. All it needed was some time.

1970s

By 1970, US President, Richard Nixon, and his Secretary of State, Henry Kissinger, had invaded Cambodia. They also waged a domestic battle: the war on drugs. The countercultural revolution had reached a crescendo. Years of sensationalist media coverage, misinformation and anti-drug propaganda had rebranded psychedelics from 'promising therapeutic medicine' to 'dangerous party drug'. Anti-LSD propaganda ran television advertisements showing teenagers jumping out of windows and displaying other forms of madness to incriminate psychedelic use. While people at risk, or with a family history of psychosis, should never take psychedelics, just like drinking alcohol, crossing the road or playing sport, psychedelics are only dangerous for those who approach it recklessly. The real danger they posed was that they could destabilise existing social systems by inspiring free-thinking individuals. Law enforcement pursued Leary and his ideals with equal fervour. In 1968, he had been arrested for having two small marijuana roaches in his car and received a ten-year sentence.[24]

In the documentary *Journeys to the Edge of Consciousness*, author Graham Hancock put it eloquently: 'The existing socially constructed reality in which we live in the 20th and 21st century tells us adamantly and clearly that our purpose of being here is to produce and consume ... We define ourselves in terms of our ability to buy more products than other people. In other words, we are defining ourselves totally in material terms in this society ... What came across to Timothy Leary was the realisation that we may be being led down a completely false path. Our purpose here on this beautiful garden of a planet and this incredible universe is not simply to buy and sell. It's not simply to define ourselves in terms of material things, but is to grow and develop the spirit. It's to develop the soul of the human being.'[25]

In the wake of the war on drugs, the US government made psychedelic compounds as restricted and unavailable as possible. They were declared Schedule I substances, a category on par with heroin and crack cocaine, and other countries followed. This classification meant that psychedelics were considered to have no medical value and a high potential for abuse. This is despite decades of genuine, groundbreaking research on their immense medical value and near zero abuse potential. Without funding for scientific studies, researchers had to abandon promising avenues of research. Even with funding, they had to navigate the red tape of the US Drug Enforcement Administration (DEA) to request an exemption, which was nearly impossible. The last fragments of interest in psychedelics went underground once again. The flame was carried by researchers and therapists who continued to work in secret, passing on their invaluable knowledge.

AFTER THE 1970s

Even though psychedelics were shunned, their influence never died. People learned to cultivate mushrooms in their bedrooms by reading manuals such as *Psilocybin: Magic Mushroom Grower's Guide* by Terence and Dennis McKenna.

Terence McKenna was called the Timothy Leary of the 1990s and equally deified. His articulation of the culture was lyrical and intellectual. 'Psychedelics are illegal not because a loving government is concerned that you may jump out of a third story window. Psychedelics are illegal because they dissolve opinion structures and culturally laid down models of behaviour and information processing. They open you up to the possibility that everything you know is wrong.'[26]

Where the values of modern society fail to inspire, dialogues by Ram Dass, Alan Watts and Jiddu Krishnamurti on spiritual philosophy endure. Classic texts such as *The Doors of Perception* by Aldous Huxley, *Food of the Gods* by Terence McKenna and *TiHKAL: A Continuation* by Alexander and Ann Shulgin line the bookshelves of psychonauts and explorers of altered states of consciousness.

Psychedelics weren't restricted to bohemian and underground tastes. Steve Jobs shared that 'taking LSD was a profound experience, one of the most important things in my life. LSD shows you that there's another side to the coin, and you can't remember it when it wears off, but you know it. It reinforced my sense of what was important – creating great things instead of making money, putting things back into the stream of history and of human consciousness as much as I could.'[27]

The aesthetics of mushrooms have been represented throughout art and culture. The essence of the mushroom is the counterculture it embodies. Mushrooms are a subtle nudge, a playful tease, a quiet rebellion against the expectations of society and government. It represents a thought tribe that prioritises freedom, love, expression and self-discovery outside the accepted guardrails we walk between for most of our lives.

Mushrooms are an ode to a time we can never return to. A time when the air pulsated with the energy of the misfits and the mavericks, the free thinkers and the free spirits, and the crazy ones who thought they could change the world. But this time it is not the counterculture. It *is* the culture.

4.2

THE SCIENTIFIC REVIVAL

Psychedelics were taboo in science for decades. It wasn't until a landmark study published in 2006 by Dr Roland Griffiths and his team at Johns Hopkins University that scientific credibility returned to psychedelic research. The study convincingly showed that psychedelics could produce remarkable effects when used correctly and that they at least deserved further investigation. Psychedelics are no longer solely seen as recreational drugs.

Griffiths, a psychopharmacologist, has studied mood-altering drugs for over forty years. His meditation practice led to his curiosity about spiritual experiences and transformation. Griffiths had heard of the experiments using psilocybin and LSD in the 1950s and 1960s that led to religious-like experiences, particularly the Good Friday Experiment conducted by the team at Harvard. But it was Robert Jesse, the founder of the non-profit Council on Spiritual Practices whose mission was to revive the science of psychedelics, who ultimately convinced Griffiths to research psychedelics.

In 1999, Roland Griffiths, Robert Jesse, William Richards, Una McCann and Mary Cosimano set off to understand the effects of psilocybin using rigorous study methodologies that were different to those of earlier studies. Their goal was to understand the immediate and long-term effects of psilocybin on healthy volunteers in a safe environment. Getting approval to mobilise this study was no easy feat. It took Griffiths and his colleagues a year to navigate the red tape with the FDA, the DEA and their institutional review board at Johns Hopkins. It became the first study in over three decades to administer a psychedelic compound to participants, sparking the beginning of the psychedelic renaissance.

The team designed a double-blind, active placebo-controlled study – the gold standard in research. The study compared psilocybin with an active control compound called methylphenidate, also known as Ritalin. Participants received 30 milligrams of psilocybin, equivalent to roughly 5 grams of dried *Psilocybe cubensis*, per 70 kilograms of body weight. The team took care to select volunteers with no prior experience with psychedelics, running pre-screening rounds and numerous questionnaires during and after the sessions. They also made sure to account for 'set and setting'. This included an aesthetic, living room-like environment that was designed for the study. During the experiments, participants were encouraged to turn inwards and were left to explore their inner self without guidance.

In 2006, the team published their landmark paper, titled 'Psilocybin can occasion mystical-type experiences having substantial and sustained personal meaning and spiritual significance' in the *Journal of Psychopharmacology*.[28] The title itself was groundbreaking. Scientific papers rarely, if ever, reference spirituality, personal meaning or mystical experiences. But the study's meticulous design, execution and data analysis commanded the attention of the scientific community.

Let's break down the title of the study. The first conclusion was that psilocybin could elicit experiences identical to natural mystical-type experiences. These are the same states that have been described by spiritual leaders, such as shamans, priests and other religious figures, throughout history. This finding suggests that spiritual experiences are biologically normal, irrespective of an individual's religious belief. Six defining characteristics of mystical experiences formed the foundation of a questionnaire. If all six criteria increased after taking psilocybin, it qualified as a mystical experience. Over 60 per cent of participants met the criteria for a complete mystical experience after taking psilocybin.

The six criteria are:

1   Feelings of unity: A strong sense of the interconnectedness of all people and things. A sense that all is one.

2   A sense of sacredness: A feeling of awe, humility and reverence.

3   Noetic quality: Knowing that what is being experienced is more real and more true than everyday reality.

4   Deeply felt positive mood: Feelings of peace, ecstasy, joy and universal love.

5   Transcendence of time and space: Loss or disruption of usual sense of time and space.

6   Ineffability: A sense that the experience could not be adequately described in words.

These mystical-type experiences are correlated with positive therapeutic benefits. Over 80 per cent said their sense of wellbeing and life satisfaction increased after the experience – no one said it decreased.

Finally, questionnaires completed by participants after the study showed that the spiritual significance of the experience persisted. There were long-term, positive personality changes, including their sense of self and life purpose. The participants' family members and friends confirmed these changes.

Griffiths remarked that throughout his research career, he had never seen anything so unique, powerful and enduring regarding the impact of a drug experience.[29] The study showed that it is possible to objectively observe the psychedelic effects of psilocybin, reuniting science and spirituality. It opened other avenues for research, including neuroscience and clinical applications for mental health illnesses such as depression, anxiety, addictions and obsessions.

4.3

PSYCHOACTIVE FUNGI

For some evolutionary reason, when we consume certain species of fungi it can trigger altered states of consciousness, including extraordinary changes in awareness, perception, cognition, emotions and mood. These changes create a psychedelic experience that transcends our physical reality, so descriptions of it are inevitably watered down. But the fact that we continue to try, through language, art and science, elucidates their spellbinding power on our psyche. This psychedelic experience is due to a set of naturally occurring compounds in fungi that are psychoactive, that is, that interact with human physiological functions. The main active compounds in psychoactive fungi are psilocybin and psilocin.

Psilocybin and psilocin can be found in psychoactive fungi throughout their life cycle, in sporing bodies, mycelium and sclerotia. Sclerotia is a mass of compacted mycelium – it's sclerotia that is sold as magic truffles in the Netherlands. Although other psychoactive fungi exist, such as *Amanita muscaria* and ergot, from which LSD is derived, psilocybin-containing mushrooms are the most prevalent in ancient cultures and in current scientific studies.

In 1958, Albert Hoffman found that psilocybin was a more stable molecule than psilocin, which made it easier to synthesise and store. Psilocin oxidises and loses its potency very quickly. This can be observed in the blue bruising that occurs when mushrooms are picked. The stability of psilocybin is why researchers use psilocybin pills in controlled studies.

Psilocybin mushrooms typically fall into the *Psilocybe* genus but are found in other genera too, such as *Gymnopilus* and *Panaeolus*. So far, over 200 species of fungi are known to contain psilocybin, and they are found across all continents except Antarctica. They are saprophytes that absorb nutrients for growth from decaying matter, such as manure and dead plant material. Their relatively simple growth cycle on a variety of substrates makes cultivation accessible to adventurous home growers. The most popular species is *Psilocybe cubensis*, colloquially known as 'shrooms' or 'magic mushrooms'.

BELOW

Psilocybe cubensis, the most famous magic mushroom.

YOUR BRAIN AND PSILOCYBIN

In the last two decades, new brain-scanning technologies have propelled neuroscience and our understanding of how psilocybin affects the brain. But our knowledge is incomplete. We do not yet understand how the changes in the brain caused by psilocybin give rise to the cascade of psychedelic effects. Here is what we know so far.

When psilocybin is ingested, the body metabolises psilocybin into psilocin. This is a crucial detail: psilocybin and psilocin are both found in the mushroom, but only psilocin acts on the brain.

Psilocin is similar in structure to serotonin, a neurotransmitter that regulates our moods. It is so similar that psilocin can hijack the receptors in the brain that are reserved for serotonin. This structural similarity may indicate a co-evolutionary development between the brain and psychedelic compounds. Psilocin can bind to a significant number of serotonin receptors in the brain, with a strong affinity for a certain type called the 5-HT2A receptor. A study showed that if the 5-HT2A receptor is blocked, no psychedelic effects occur.[30] Therefore, the activation of these receptors appears to be critical for initiating psychedelic effects. However, it is still not known how the activation of serotonin receptor sites creates the psychedelic experience.

A team, led by Dr Robin Carhart-Harris and Professor David Nutt from Imperial College London and funded by the Beckley Foundation, investigated the effects of psilocybin on the brain in 2012. They used functional magnetic resonance imaging to scan the brains of volunteers on psilocybin. Using blood flow as a measure of brain activity, the team found that activity decreased in a part of the brain called the default mode network.[31] Interestingly, this area is also home to a large proportion of 5-HT2A receptors.

The default mode network is an interconnected group of brain regions that play a central role in brain functions, like a conductor in an orchestra. This network also plays an important role in consciousness. It is associated with metacognitive functions, such as the ability to time travel (i.e. to think about past and future events), contemplate ourselves with introspective awareness, and empathise with other people's perspectives. This part of our brain helps us form our identity, allowing us to define who we are and how that is projected to the outside world. Loosely put, the default mode network is where our ego lives – the feeling of our individual identity. It is also where critical self-talk takes place. When this part of our brain is overactive, thought and behaviour patterns can loop, which may lead to excessive rumination. Rumination

Psilocybin and psilocin have chemical structures that are remarkably similar to serotonin.

Psilocybin

Psilocin

Serotonin

has been shown to exacerbate negative emotions and deepen depression and anxiety. If unchecked over time, this pattern of thinking becomes rigid and is hard to break out of.[32]

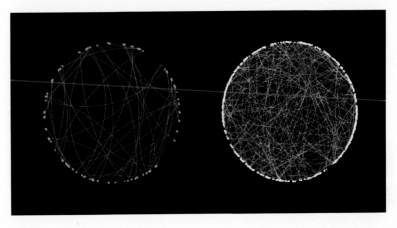

When the activity in the default mode network quietens, it releases its grip as the central conductor. Personal narratives recede, allowing an escape from habitual thinking and defences. When the rigid boundaries of the mind dissolve, the sense of self can diminish or disappear, which is called ego dissolution. Another effect of a muted default mode network is that other parts of the brain that don't normally communicate with each other start to form new connections.

One such phenomenon is synesthesia, when the senses seem to blend together. For example, when the gustatory cortex speaks to the visual cortex, you can 'taste' colours. This hyperconnectivity facilitates unconstrained cognition, spontaneous insight and creativity.[33] In this open state, the brain is more flexible, so there is a chance to rewire it. Freedom from the ego creates a space that can allow repressed emotions or states of bliss to surface. When the curtain lifts, a constellation of new realisations, perspectives and feelings about life punctuates the experience. This can create lasting personal transformations if these learnings are integrated into daily life.

Ego dissolution is temporary. The default mode network is necessary for us to function in our complex society, and it starts up again as the psilocybin experience winds down. The window of opportunity is only open for about six hours, but when used intentionally this brief period of time can provide insights that last a lifetime. Dr Stanislav Grof, psychiatrist and author of *The Way of the Psychonaut*, believes that 'Psychedelics, used responsibly and with proper caution, would be for psychiatry what the microscope is for biology or the telescope is for astronomy.'[34]

There is a proven relationship between psilocin, serotonin receptors and the default mode network, but how consciousness is actually produced and altered is still a mystery. Indeed, consciousness is only recently starting to become an accepted medical research subject. This is despite the fact that all we are, and all that we will ever experience, unfolds in consciousness. How consciousness is borne from an assembly of neurons in the brain is still far from being understood. Unless we are asking the wrong questions – does consciousness operate independently of the body? Whatever the answer is, consciousness is built into our subjective experience of life. Our thoughts revolve around a sense of self, or the ego, supplemented by information from our senses and deep layers of conscious and unconscious beliefs. We have the capacity to appreciate, laugh and connect, just as we can be riddled with anxiety, confusion and worry. These are all our own experiences, distinct from the rest of the universe.

4.5

PSILOCYBIN–ASSISTED THERAPY

Psilocybin-assisted therapy involves using psilocybin in conjunction with a form of talk-therapy to heal underlying psychological conditions. The therapeutic approach is tailored to the participant and their condition. Psilocybin-assisted therapy has three stages, all of which are critical for the therapy's effectiveness and safety.

Preparatory sessions between the guide and participant establish trust and rapport, which are critical during the psilocybin experience. A safe environment allows the participants to be vulnerable and share intimate details about their life story. These details often come up during the psilocybin experience. The guide will also prepare the participants by informing them about what to expect, and how to navigate challenging experiences if they arise.

During the psilocybin session, the participant lies down in a safe, comfortable environment that has been specifically designed for them. They are encouraged to go inwards with the aid of an eye mask and headphones that play a curated music playlist. The guide is there for reassurance and support in challenging situations. In this environment, the body's innate self-healing ability is given the space to switch on. Just as the body knows to heal an open wound, it can guide the participants to areas of the psyche that need healing.

The end of the psilocybin session is when the work starts. With support from their guide, the participant discusses their experience to unpack the insights from their sessions. Without integration, this becomes just another experience. It takes time and dedicated practice to embody the new learnings and positive behaviours.

When used safely, psilocybin can dissolve the lines that we draw around ourselves in times of crisis. It can unlock deep traumas. These experiences need to be held in a safe container with a trusted guide so that the insights from the session can be integrated into daily life moving forward.

The teachings of the psilocybin mushroom

MARY COSIMANO

Mary Cosimano, LMSW, is currently with the Department of Psychiatry and Behavioral Sciences at the Johns Hopkins University School of Medicine, one of the world-leading centres conducting human research with psychedelics. She is the Director of Guide Services for the Center for Psychedelic and Consciousness Research and has served as study guide and research coordinator for psilocybin studies for twenty years. During that time, Mary has been a session guide with psilocybin and club drug studies, and has conducted over 450 study sessions. She teaches at California Institute of Integral Studies in their Psychedelic-Assisted Therapies and Research program and also conducts training for therapists in psychedelic psychotherapy.

It was during our fifth psilocybin study, 'Effects of psilocybin and spiritual practice on persisting changes in attitudes and behavior', at Johns Hopkins that I began to understand the teachings of the psilocybin mushroom on a deeper level. The purpose of this study was to investigate how psilocybin, when combined with spiritual practices such as meditation, could affect the behaviour and attitudes of participants. Participants were given questionnaires that measured the level of mystical experience and states of consciousness they encountered during the psilocybin session, followed by a persisting effects questionnaire three weeks later.

In 2011, approximately two years into this study, I was conducting the psychological assessments, a pre-screening process that was mandatory for participants to enter the study. During many conversations, I began noticing the difficulty and disconnect many participants were experiencing due to the word 'spiritual', which was present in the name of the study. This word often has significant baggage associated with it, as does the word 'religious'. I felt a need to make sense of this, because I believed that, deep down, the essence of our studies are well beyond either of these words.

I realised the phrase 'authentic or true self' was more accurate than the word 'spiritual'. This seemed to resonate with many participants, as it was less confusing. I came to believe that the power of the psilocybin mushroom was that it seemed to act as a key that opens our minds and allows access to the part of ourselves that is our 'true self'. This 'true self' can emerge when the blocks or barriers we have constructed have been removed.

I began to see that what we describe as our authentic selves are our true nature. And that our true nature is love, which I believe is our connection, our relationship – to ourselves, to others and to everything. Deep down it is what we all want, to love and be loved. It is our true essence and who we are. We came into the world knowing this, but as we grow, we often become disconnected from this knowledge. We make our lives so complicated. Then we develop fears and, because of our fears, we put up walls and barriers that disconnect us, and thus we lose our sense of who we are – our true authentic self. I believe our psilocybin studies are about reconnecting to our authentic or true self no matter the age, socioeconomic status or cultural background. When someone experiences this it is often a familiar feeling, comfortable, like coming home. As one participant wrote of their experience, 'The whole journey was on the spiritual plane where thoughts, understanding and words are completely, 100 per cent irrelevant. I was in the playground of trust and knowing. I was free from the binds of my overactive mind. I was free from density. I was in my element. This is my home. This is what I know is my true self.'

This conviction – love and connection to our true, authentic self – addresses what I believe is one of the main outcomes of our psilocybin studies. As one participant put it: 'Everything is swept up into a climactic epiphany of love as the universal essence and meaning of all things.' There are literally hundreds of quotes from participants' psilocybin session experiences that speak to this same idea.

The teaching of the psilocybin mushrooms are many, and I noted recurring themes based on my twenty years of witnessing psilocybin sessions, participants' session reports and the observations of their friends and colleagues. The core teaching, as I said, is an increase in a core sense of self, a deep-rooted knowingness of our true, authentic self. And an interconnectedness to all of life – a knowing that we are all one, all connected.

The result of knowing who we truly are translates into a desire to be a better person – to ourselves, to others, and to the world. Participants have reported significant increases in quality of life and personal meaning beyond the psilocybin experience. They have shared feelings of deep gratitude and the importance of forgiveness and compassion for self and all, and priorities shifting – awareness of what's important and letting go of what's not and decreasing petty concerns and holding onto grudges.

The following is a beautiful summary from one of our participants. 'I am more interested in connecting with others on a deeper level, and value relationships more. I am more open and less judgemental. I have a greater sense of love as being a unifying and primal force in the universe and as the appropriate response to my understanding of reality.'

Finally, I would like to share how these past twenty years have impacted me personally. On one hand, I could write volumes, yet on the other, there is no way to put it into words – it is ineffable. Thus, I will end on this note. Being an integral part of our participants' lives and their experiences these past two decades has deepened my existing belief that love is everything to a level I'd not known before. Globally, I wish for these medicines to be available to all those who are ready for it, meaning they are medically and psychologically safe to receive it and the potential value it can have for them and the world.

THE PSILOCYBIN EXPERIENCE

NOTE: This is not advice or an instruction guide for consuming psilocybin mushrooms. This is a description of what a psychedelic experience might entail.

PREPARATION

In their natural form, psilocybin-containing mushrooms can vary widely in their concentration of psychoactive compounds, even within the same species and harvest location. That said, they are still generally dosed by weight. A rule of thumb is that fresh mushrooms weigh ten times more than dried mushrooms, due to their high water content.

Mushrooms can be consumed fresh or dried in any number of ways. They can be blended into a drink, steeped in hot water to make a tea, chewed or cooked as food. Depending on the consumption method, it can take twenty to sixty minutes for the psychedelic effects to begin.

Traditional rituals include a fast of at least six hours before consuming mushrooms. An empty stomach can improve the spiritual experience and help avoid nausea, which is a possible side effect. Consuming mushrooms without fasting can delay the onset of the psychedelic effects.

There is no 'typical' psilocybin experience. There are many variables involved but optimising 'set and setting' will positively impact the experience. It's normal to be nervous, but trust the process. Surrender to what comes next, however it comes.

DOSAGE

Psilocybin mushrooms have varying effects for each individual and at each dosage level, making the experience different for everyone. Below is an exploration of what may occur. The experience will often last between four and eight hours, depending on the potency of the mushrooms and the dose consumed.

NOTE: All measurements are based on dried *Psilocybe cubensis* mushrooms.

Microdose (0.1–0.25 g)

A microdose is a sub-perceptual dose where the psychedelic effects are not noticeable. It does not affect your visual and auditory perception. At this level, everything can feel simply a little bit better. It can add a shine to your day and enhance clarity, focus, energy and creativity. This is why microdosing is popular within Silicon Valley and the wider technology industry. Users report that they can access a flow state, which combines intellectual stamina and spontaneous inspiration.

Low dose (0.25–1 g)

This is a 'museum dose', a term coined by chemist and godfather of psychedelics Dr Alexander Shulgin. This light dose paints your environment with a layer of sparkle and awe. Some light psychedelic effects may be noticed, but your ability to function and participate in social activities is still intact. This is also called a concert dose, as it is used recreationally.

Moderate dose (1–2.5 g)

This dose is popular for creative problem-solving. The mind relaxes and allows new ideas, insights and connections to flow in. At this level, the senses are enhanced with mild visuals. Social barriers dissolve, so this dose can help you reconnect with loved ones and nature.

High dose (2.5–5 g)

At this dose, the full psychedelic experience will start to come on with geometric patterns and fractals. Time and space may become warped. Cognitive tasks become more difficult. You can still grasp your surroundings, but it starts to become altered and distorted.

A high dose is introspective and can be used as a tool for inner exploration and healing. As you move into a higher state of consciousness, the everyday voice of the ego quietens. You can objectively see yourself and ask questions such as 'Who am I?' and 'What is important to me?' Concepts

of the self that you attach yourself to can start to disappear as your ego dissolves. The experience may be overwhelming or confusing, but the fire of the psychedelic experience does not burn you – it only burns what you are not.

Heroic dose (>5 g)

The heroic dose was famously created by Terence McKenna. He instructed the bravest of psychonauts to consume 5 grams of dried mushrooms alone, in the dark, with their eyes closed. He said 'If you take a psychedelic and you're not afraid you did too much, you didn't do enough'.[36] This is a committed dose where the ego can cease to exist, leading to a total loss of self-identity referred to as ego death. There can be an awareness of being connected to all people and the whole universe, and a complete transcendence of the self – all characteristics of a mystical experience.

As the dose increases dramatically over 5 grams, it is possible to enter a void-like state where words are insufficient to explain the experience. It can be difficult to bring back any insights from the void. Proceed with caution.

THE ENTRY

'Coming up' into the psychedelic experience, you may feel tingling in the body, a need to giggle and a sense of relaxation or nervousness. Perhaps the best way to describe this is 'not normal'. Your senses may become enhanced. You might see brighter colours, solid objects becoming wavy or plants that seem to be breathing. Concepts such as time and space can melt away.

THE TRIP

These changes intensify, and the visual effects can become fractal geometries or kaleidoscopic patterns, especially if your eyes are shut. Try closing your eyes – there is a whole universe inside of you.

The psychedelic experience also affects your thoughts and feelings. Mushrooms can shine a light on what is already inside you. The conscious and subconscious parts of your psyche reveal themselves. Repressed traumas, fears and memories can surface. If you allow it, you can expand your awareness and be taken to incredible places to explore the complexity of the human condition. This can be overwhelming and challenging, but surrendering to whatever may arise is critical to healing. When you do surrender, a feeling of love, peace and familiarity can take over. You may feel a sense of connection to others, nature and the cosmos. Welcome home.

ON BAD TRIPS

Psychedelics are one of the safest drugs. In a study of twenty drugs, the UK's Independent Scientific Committee on Drugs ranked psilocybin as least harmful and LSD as the third-least harmful. Alcohol was ranked the most harmful – more than ten times as harmful as psilocybin and LSD – followed by heroin and crack cocaine.[37]

Psychedelics are not addictive and do not harm the body or brain.[38] But there are some people who should never take psychedelics, including those with a personal or family history of psychotic disorders. Psychedelics can also affect blood pressure and heart rate, so people with heart conditions may also be at risk. Psilocybin-assisted therapy uses extensive questionnaires to determine the suitability of a participant before administering psilocybin.

Proper preparation is the deciding factor between a challenging experience – one from which you can grow and integrate – and a bad trip. A bad trip is characterised by feelings of anxiety, fear, paranoia, panic and resistance to whatever is unfolding. Because emotions are intensified during a psychedelic experience, new feelings, thoughts or experiences that arise can be difficult to manage if the person is underprepared or inexperienced.

Set and setting are the key determinants to a positive experience. Understand what your intention is and commit to accepting what the psychedelics show you. Ask a trusted friend or family member to sit with you for support. Optimise your surroundings – make it safe, comfortable and incorporate ritualistic elements, such as incense, music or candles. Don't mix psychedelics with other drugs or alcohol.

Psychedelics show you what you need to see, including repressed thoughts and trauma, or highlight characteristics of yourself that you have never seen. There's a lot of unravelling, unlearning and letting go. But it all starts with acceptance. Resistance is what causes a bad trip. Remember that the trip always ends. Integration is the key to all trips, especially a challenging one. What affects you deeply during a trip is probably something in your psyche that you have not resolved. It is best if this is met with further inquiry and compassion.

THE INTEGRATION

The real journey starts here. What you saw, learned and experienced needs to be brought back to everyday reality and embodied. Not just for your own healing, but for the collective healing of the world. Society desperately needs these new ideas and perspectives. As the prolific author and speaker Alan Watts said, 'When you get the message, hang up the phone. For psychedelic drugs are simply instruments, like microscopes, telescopes, and telephones. The biologist does not sit with eye permanently glued to the microscope; he goes away and works on what he has seen.'[39]

Mushrooms can shine a light on what is already inside you. The conscious and subconscious parts of your psyche reveal themselves.

PREPARING FOR A PSYCHEDELIC JOURNEY

Before embarking on a psychedelic journey, optimise your set and setting for a positive experience. 'Set' refers to the mindset of the person taking the psychedelic. 'Setting' refers to the physical and social environment. All current scientific studies involving psilocybin-assisted therapy optimise 'set and setting' for their participants.

THE RIGHT SET

Have clear intentions, be in a good headspace and respect the sacred medicine. It is normal to feel nervous, anxious or excited, but going into a psychedelic journey in a state of equanimity will help ease you into the experience. Activities such as meditation, yoga, deep breathing and spending time in nature can also help you relax.

Ask yourself what your intention is for entering this journey, even if it's just to explore. Observe your inner world and be aware of the state you are in. Psychedelics can amplify your existing feelings, so assess your emotional state and check that you are comfortable entering a psychedelic experience.

Let the experience take you where it wants to. Excessive control and expectations can create tension. Be ready to surrender to what arises.

THE RIGHT SETTING

Plan your social and physical environment. Ensure that there will be no disruptions and don't make plans for the rest of the day. You can journey alone, with a guide or with trusted companions, but make sure you are in a safe and comfortable environment. If it's your first time, a sober guide can provide additional psychological safety.

The setting can be tailored to the type of psychedelic experience you seek and the dose you take. For an inner journey, staying indoors or at home is a good option, as this is often where we feel most grounded and safe. A journey in nature can also be a special experience to reconnect with the natural world. Ensure the outdoor environment is away from unwanted noise and stimuli. This will help create a more peaceful journey.

4.6

THE PSYCHEDELIC RENAISSANCE

Since the 2006 landmark study by Griffiths and his colleagues, the FDA has approved over fifty clinical research studies using psilocybin. The idea that psychedelics can play a huge role in alleviating human suffering is now a legitimate scientific research area. New studies looking to treat depression, end-of-life anxiety, post-traumatic stress disorder, anorexia, alcoholism, smoking addiction, Alzheimer's disease and a host of other conditions are underway. The applications are seemingly infinite, although researchers, heeding the mistakes of the 1960s, are cautious. Research efforts are being led by renowned universities such as Johns Hopkins University, New York University, Imperial College London and University of California, Berkeley, all of whom have opened centres dedicated to psychedelic research.

The study of consciousness and psychedelics is also becoming an accepted course of study for graduate students. Areas of research range from studies into the effects of microdosing psychedelics, the long-term effects of psychedelics on the brain, future design of psychedelic-assisted therapies and even activities outside medicine, such as anthropology, art, music, computer science, philosophy and religion.

Psychedelic research has returned during a deepening mental health crisis. The statistics are truly sobering: over 1 billion people worldwide now live with a mental health issue, with depression being the most significant factor for suicides. In the US, depression among adults more than tripled between 2018 and 2021.[40] In the UK, 30 per cent of patients with depression are resistant to treatment. The world urgently needs new solutions. The current treatment options for depression include psychotherapy, usually supplemented with a daily antidepressant called a selective serotonin reuptake inhibitor (SSRI). SSRIs work by muting the brain's stress system, but this doesn't address the root cause of depression and can cause a host of side effects.

Dr Carhart-Harris and his team at Imperial College London are adding to the growing body of research proving that psilocybin-assisted therapy can treat depression. In April 2021, they published the results of a phase II double-blind, randomised, placebo-controlled trial in *The New England Journal of Medicine* called 'Trial of Psilocybin versus Escitalopram for Depression'.[41] This six-week study recruited fifty-nine participants who had mild to severe depression. The goal was to compare the efficacy of two treatment options over two months: two sessions of psilocybin-assisted therapy against a daily dose of escitalopram, an SSRI widely considered the best antidepressant on the market. The escitalopram group also received the same form of psychotherapy as the psilocybin-assisted therapy group.

The researchers calculated depression scores for each participant and the average depression score for each group. At the end of six weeks, the average depression score in both groups reduced without significant differences, showing that two doses of psilocybin-assisted therapy are at least as effective as sixty-three daily doses of escitalopram. That may seem immaterial, but the psilocybin group reduced their depression scores faster. Also, 70 per cent of the participants in the psilocybin group at least halved their depression scores, compared with 48 per cent of participants in the escitalopram group. Of the people who experienced a reduction in depression symptoms, twice as many maintained it at the end of six weeks in the psilocybin group than those in the escitalopram group. These results are extremely promising.

Dr Carhart-Harris believes that participants who experienced the psilocybin-assisted therapy sessions were given an opportunity to liberate themselves from the imprisonment of their ruminating minds. Once free from their obsessive, dominating stories, they were able to unearth repressed thoughts and emotions and work through them more objectively. In contrast, escitalopram and other SSRIs do not address the root causes of depression. They level out a person's depression symptoms by inhibiting responses in the brain. Dr Carhart-Harris hopes to conduct a longer-term study in other sites on a more diverse population of participants. If successful, those with depression may have a new treatment option in the near future that requires fewer doses and has fewer side effects.

Scientific researchers aren't the only ones interested in the potential of psychedelics. Mainstream media, regulators, companies and individual advocates across many fields are excited about the future of psychedelics, too. The attitudes of regulatory bodies are also changing. In the US, the FDA advanced psilocybin studies into phase II trials. In 2018, it granted 'breakthrough therapy' status to a psilocybin-assisted therapy for treatment-resistant depression run by Compass, a life sciences company. This means that the FDA has placed clinical significance on the preliminary results of the trial's efficacy. The FDA will fast-track its review process for the final phases of clinical trials, likely making Compass the first in line to bring psilocybin-assisted therapy to market. Compass became a publicly listed company on NASDAQ in September 2020 and is the first psychedelics company to reach a market capitalisation of over US$1 billion.

In the last few years, there has been an influx of companies entering the psychedelics industry. ATAI Life Sciences, MindMed, Numinus, and Field Trip have also listed on the public market. They aim to raise capital to complete research, development and testing, alongside regulatory approval of novel compounds and medicine delivery methods. Marshall Tyler, the Director of Research at Field Trip, explained that they have developed a new synthetic molecule called FT-104, which mimics psilocybin. 'With psilocybin, someone tends to be in the clinic for six to eight hours, which is incredibly cumbersome for the patient and costly for the clinician. If you can get someone in and out a lot quicker than that, but have a similar intensity of experience and therapeutic benefit, that's something that's positive and one of the focuses with FT-104.' [42] Pre-clinical trials have demonstrated the potential of FT-104 that Tyler describes.

With more companies creating novel compounds based on psilocybin, debates concerning the ethics of patenting a naturally occurring compound – however modified – are developing. Patents have also been filed for entirely new psychedelic compounds and methods of psychedelic-assisted therapy. Of course, investment does fund research and development, build clinical infrastructure and fast-track commercialisation models to distribute medicine to those in need. But concerns have been raised about the potential monopolisation of psychedelic medicines by for-profit companies that are largely driven by shareholder returns.

This leads to more questions about how we can ensure equal access to these medicines and how they will be administered. The current medical model only allows patients with an approved diagnosis to be treated within the medical system. This neglects those who don't fit the prescribed indications, and further, the cost of this access route may be prohibitive. Low socioeconomic groups have a higher risk factor for mental illness, and policymakers need to prioritise their access to these and other potentially revolutionary treatments.[43]

These are necessary debates as we create the first framework for psychedelic medicine in the Western world. Let's not forget that psychedelic research and therapies owe much of their success to the traditions of Indigenous groups around the world. It is essential to acknowledge these custodians and their healing practices, which we borrow and base our studies on. Indigenous groups need to be allowed to move out of the marginalised sidelines and be given a say about the development of a new healthcare system. For those who are willing to align their work in the field of psychedelics to a set of guiding values, the pledge developed by the advocacy group North Star is the first step. The North Star Ethics Pledge involves committing to a set of values, with additional education and actionable steps to ensure psychedelic wisdom and integrity is woven into the growing psychedelic industry.[44]

4.7

DECRIMINALISATION AND THE ROAD TO LEGALISATION

The Schedule I classification of many psychedelics was more likely due to politics in the 1970s than the scientific potential of the compounds. Throughout the psychedelic 'dark ages' that followed, the non-profit Multidisciplinary Association for Psychedelic Studies (MAPS) was the leading force for scientific research and education on safe and beneficial uses of psychedelics. Its founder, Rick Doblin, was an undergraduate at Harvard during Leary's Harvard Psilocybin Project days, and dedicated his career to furthering psychedelics in medical and legal contexts by combining drug development with drug policy reform.

After thirty-five years, MAPS is close to achieving FDA approval for MDMA-assisted psychotherapy for post-traumatic stress disorder. It is in phase III trials, the furthest level of clinical development in the US for any psychoactive substance. As an advocate for the healing potential of all types of psychedelics, Doblin believes that 'Psychedelics must be legalised and regulated to some extent to permit therapeutic applications to occur without fear of prosecution'.[45] As we enter a psychedelic society, drug reforms across the world are unfolding.

In May 2019, Denver became the first city in the US to decriminalise 'the use and possession of mushrooms containing the psychedelic compound psilocybin'.[46] In quick succession, Oakland, Washington DC, Ann Arbor, Santa Cruz and the state of Oregon also moved to lower the penalties associated with personal use and cultivation of psilocybin mushrooms. Decriminalisation means that psychedelics are still illegal, but it will be the lowest priority for law enforcement. Many cities and states are expected to follow suit.

In November 2020, Oregon became the first US state to legalise psilocybin for therapeutic use. Oregon's Measure 109 allows licensed service providers to administer psilocybin-containing fungi products to adults over twenty-one years of age in therapeutic settings under the guidance of trained professionals. These professionals do not need to be medical doctors but will need to undergo training to help guide psychedelic experiences from preparation through the trip itself and integration afterwards. Adults will not need to get a prescription or visit a medical setting to access psilocybin. Psilocybin remains illegal outside of these licensed service centres.

In June 2021, the California State Senate passed Senate Bill 519, a legislation that decriminalises psychedelics, including psilocybin and LSD. The bill needs to be passed by the California Assembly before it becomes law. If this happens, then criminal penalties for growing, possessing and sharing psychedelics will be removed for adults over twenty-one years of age, effectively legalising psychedelics. California was a pioneering state in the legalisation of medicinal cannabis in 1996 and is poised to lead the US – and the world – into psychedelic legalisation and a new era of drug policy reform.

We can expect psychedelic companies to enter regions that implement a form of legalisation. In other countries, such as Canada, the UK and Australia, there is also a trend towards decriminalisation and legalisation being led by grassroots activists.

As the psychedelic future gains momentum, so too does the underground network of psychedelics. With a vast amount of resources available from websites such as Shroomery and Erowid, people can source, grow and consume psychedelics in their homes. A hidden industry also exists for professional guides to facilitate the use of psychedelics in a safe and supportive environment for individuals who have limited treatment options. Alternatively, countries such as Jamaica and the Netherlands offer a form of legal psilocybin. Synthesis, an organisation based in Amsterdam, offers a three- or five-day medically supervised psilocybin truffle retreat. However, for those who have fears of persecution or cannot afford to travel, their hopes remain tethered to the legalisation of such compounds in their local regions.

Cultural commentators drive the dissemination of psychedelic news and education. Using their expertise and varied communication platforms, they are shifting the public perception of psychedelics. Michael Pollan is the author of the *New York Times* bestseller *How to Change Your Mind: What the New Science of Psychedelics Teaches Us About Consciousness, Dying, Addiction, Depression, and Transcendence*. Published in 2018, Pollan's book turned the tide for psychedelics, reaching a mainstream audience with his investigative journalism interspersed with scientific research and accounts of his own psychedelic experiences.

Tim Ferriss, author, investor and host of *The Tim Ferriss Show* podcast, is an important voice and influential advocate for the psychedelic movement. With a family history of treatment-resistant depression, bipolar disorder and addiction, he has witnessed and experienced firsthand the healing potential of psychedelic medicine. Ferriss was a key fundraiser for the new Johns Hopkins Center for Psychedelic and Consciousness Research, personally pledging

US$2 million. Ferriss also lends his voice to the sustainable harvest and use of psychedelic plants and fungi. He advises: 'Choose species that grow well, grow widely, and grow quickly,' encouraging a focus on 'readily available and easily cultivated species like *Psilocybe* mushrooms ... [which] can take weeks to grow'. 'If you could only use *Psilocybe* mushrooms for the rest of your life,' he says, 'you could continue to cultivate that relationship, develop deep skills, and unfurl profound layers of learning and meaning until your dying breath. The depth is there, if you commit to the exploration. There is no need to stamp the psychedelic passport with every plant or animal, and there are many reasons not to.'[47]

As we move toward a psychedelic society, we hope that psychedelics and the sacred rituals they stem from are respected. Our ancestors approached these natural medicines with reverence, love and humility, and we must do the same. The psilocybin mushroom should be available to anyone brave and curious enough to explore the fabric of this world. The road to legalisation and integration into society will be challenging, but we need to take risks for the chance at a more healthy, connected and conscious society.

Our ancestors approached these natural medicines with reverence, love and humility, and we must do the same.

4.8

REDISCOVERING PARADISE

The power of psychedelics goes beyond therapeutic applications. As scientific studies have proven, psychedelics can reliably induce mystical states of consciousness, suggesting that we are biologically wired to have these experiences. But what are altered states of consciousness? More importantly, what can we learn from entering higher states of consciousness?

Consciousness is a mysterious phenomena, so the term has different definitions depending on the context. One way to understand consciousness is to see it as the medium through which we experience the subjective world around and within us. Our level of consciousness is measured by our level of awareness and attention. This determines the quality of our experience of life and how 'truly alive' we feel. Our level of consciousness changes throughout the day. In deep sleep, all conscious thoughts and emotions are shut off – there is no awareness. The dreaming stage involves a string of images, thoughts and sensory experiences, but our level of awareness of what is happening is low. When we are awake, our levels of awareness also range. Most of us are in a waking sleep, doing tasks mindlessly while our attention is elsewhere. Think about the hours spent caught up in our own narrative rather than being in the present moment. To be fully awake and present requires focus and practice.

Consciousness is always total and complete – it merely appears clouded and foggy because of our attachments, labels and other illusions. For thousands of years, Eastern philosophies such as Hinduism and Buddhism have shared the practice of mindfulness to help bring our wandering minds back into the present moment. During these times of quality presence, we can reach a higher state of consciousness. These moments allow us to witness the nature of our experience with full clarity, without being captured by thoughts of the past or future. Allowing such states of consciousness to punctuate your day is not an abstract idea, nor is it confined to spirituality or religion.

In 1943, psychologist Abraham Maslow created the seminal hierarchy of needs in a paper called 'Theory of Human Motivation'. He was trying to understand what motivates human behaviours. It goes like this. Firstly, humans need shelter, food, water and rest for survival. Next, we need 'safety', defined as money, security and lack of fear. Then we seek love and belonging in a community. Then we craft our individual identity with achievements and respect from others. Once these deficiency needs are fulfilled, we seek self-actualisation. 'A musician must make music, an artist must paint, a poet must write if he is to be ultimately happy,' Maslow said. 'What a man can be, he must be.'[48] But he wasn't finished.

Before Maslow's death in 1970, he was working on the concept of peak experiences – the final step after self-actualisation. Humans need to grow beyond our individual identities. We need to transcend ourselves, explore humanity and love nature and others. A person has peak experiences when they are in harmony with themselves and their surroundings, and these experiences are characterised by states of ecstasy, wonder and joy – the same qualities of higher states of consciousness.

'In the last few years it has become quite clear that certain drugs called "psychedelics", especially LSD and psilocybin, give us some possibility of control in this realm of peak-experiences,' said Maslow in his 1964 book *Religions, Values, and Peak-Experiences*. 'It looks as if these drugs often produce 'peak-experiences' in the right people under the right circumstances, so that perhaps we needn't wait for them to occur by good fortune.'[49] Psilocybin, even in moderate doses of 3 to 4 grams, tends to dissolve our attachment to deficiency needs. Maslow chose the term 'peak-experiences' to remove the religious connotations, so that it was understood that anyone could achieve these egoless moments of self-transcendence.

Self-transcendence occurs at higher states of consciousness, when we are no longer violently battered by our mind's anxieties, preoccupations, fears and impulses. Instead, the self becomes small and the sense of connection to a cause, an idea or a collective that's bigger than ourselves arises. It's a liberation. In moments of self-transcendence, we realise that we aren't lonely fragments floating through time, but that we are all expressions of nature, connected with and to each other by spirit. 'We do not "come into" this world,' said Alan Watts, 'we come out of it, as leaves from a tree. As the ocean "waves", the universe "peoples". Every individual is an expression of the whole realm of nature, a unique action of the total universe. This fact is rarely, if ever, experienced by most individuals. Even those who know it to be true in theory do not sense or feel it, but continue to be aware of themselves as isolated "egos" inside bags of skin.'[50]

Psychedelics can reliably occasion these experiences of unity. But so can many other experiences – yoga, meditation and time in nature, just to name a few. It is when we no longer project a distinct 'me' and 'other' – that is, when consciousness is no longer divided into subject and object – that wholeness can arise. Reality is undivided wholeness. Poets, artists, writers, athletes and performers call this 'flow'. It taps into something very old, very wise and very deep within us. When you wake up to the fact that you are whole with life, life starts to move through you. When we wake up to the realisation that we are amid the most wonderful miracle of life itself, we remember that this is, and always has been, paradise.

Why are we biologically wired to have these experiences of higher states of consciousness? Is there a spiritual instinct in all of us? Is there an innate compass guiding us? Whatever the answers to the mysteries are, we have been endowed with the ability to make sense of them. It is up to you to explore it.

The world is awakening to a new consciousness, and psychedelics are playing a vital role. Most cultures throughout history have had intimate relationships with psychedelics to alter their consciousness. Fungi remind us that we are stewards of our consciousness and of the planet. It is from this place of connection that we can begin to heal. This is how shamans have guided their communities for millennia. Psychedelics guided them, and psychedelics can guide us too.

We are all expressions of nature, connected with and to each other by spirit.

AMANITA MUSCARIA

COMMON NAMES

Fly agaric, fly amanita, *beni-tengu-take* ('red long-nosed goblin mushroom' in Japanese), *mukhomor* ('fly killer' in Russian), *tue mouche* ('fly killer' in French)

FAMILY	Amanitaceae
GENUS	Amanita
SPECIES EPITHET	muscaria

Arguably the most iconic and recognisable mushroom species, *Amanita muscaria* is a brilliant red or orange mushroom with white patches. The spots are remnants of the white veil that enclosed the mushroom when it was young. Its common name is 'fly agaric' because it was traditionally used to attract and kill flies. The cap was crushed in a saucer of milk and set as a trap on windowsills. It is widely represented in popular culture, including appearances in *Super Mario Bros.*, *Alice in Wonderland* and *The Smurfs* – and the classic mushroom emoji.

HISTORY AND CULTURE

Many myths and legends surround the use of *A. muscaria*. For hundreds of years, it has been used as an entheogen in religious contexts to reach trance-like states. However, there is a fine line between psychoactive and toxic. In larger doses, it can cause sweating, twitching, nausea and diarrhoea. Boiling and drying can decrease the toxicity without compromising the psychoactive effects. Another (riskier) method involves using the human liver to filter out the toxins. This is commonly practised by Siberian shamans – they ingest the mushrooms so that their livers filter out the toxins, leaving the psychoactive compounds in their urine for others to consume.

PROPERTIES

EDIBLE

Yes, but not recommended for consumption in raw form due to intoxicating properties. Contains muscimol and ibotenic acid, which can be dangerous in large amounts, although fatalities are extremely rare. Edible if boiled in water, which weakens its toxicity.

MEDICINAL

Yes. Rich history of traditional use in Siberia, Russia, and eastern and northern Europe. Used topically to treat muscle and joint pain, tissue injuries and post-workout soreness.

PSYCHOACTIVE

Yes. Contains muscimol and ibotenic acid, which can produce effects of a 'waking dream' such as delirium, detachment, dizziness, stillness and clarity of perception. This is different from the effects of psilocybin-containing mushrooms.

ENVIRONMENTAL REMEDIATION

Yes. Shown to accumulate metals such as mercury, copper and zinc from forest soils into its sporing body.[51]

SPORING BODY CHARACTERISTICS

CAP

5–25 cm wide
Flat or convex
Bright red or orange to yellow
Dotted with raised white warts

GILLS

White
Close or crowded
Free or nearly free from stipe

STIPE

5–20 cm tall
1–3 cm thick
Bulbous volva at base
White to yellow-white
Smooth or scaly
Off-white upper ring, may be toothed

SPORES

White
Oval

FIELD DESCRIPTION

HABITAT

Commonly found in mycorrhizal relationships with trees, particularly pine, spruce and birch. Often in groups or rings in the soil. In Australia, it allies itself with eucalypts, which is in contrast to its usual habitats of snowy pines in Europe and Asia.

DISTRIBUTION RANGE

Found throughout North America, Europe, Asia and Australia.

SEASON

Summer and autumn.

PSILOCYBE CUBENSIS

COMMON NAMES

Magic mushroom, shroom, gold cap, cube

FAMILY	Hymenogastraceae
GENUS	Psilocybe
SPECIES EPITHET	cubensis

Cubensis is Latin for 'from Cuba', which is where *Psilocybe cubensis* was first collected in 1904. Its cap is golden-brown with white flecks and has a tendency to stain blue when touched. The mushroom behind the magic is a dung-loving species that stays close to grazing cattle. It consumes the nutrients in the dung and creates the sporing body to release spores that in turn germinate in other dung piles.

HISTORY AND CULTURE

P. cubensis is the most cultivated psychedelic mushroom in the world, particularly the strain 'golden teacher'. Guided by the McKenna brothers' book *Psilocybin: Magic Mushroom Grower's Guide*, travellers who went to Central and South America to find these mushrooms and brought back spores could cultivate their own at home. It's not a picky species – with limited equipment, zero prior experience and eight weeks of love and patience, a flush of *P. cubensis* can grow in any dark, empty cupboard.

PROPERTIES

EDIBLE

Yes, but it is psychoactive. Not recommended for a family dinner.

MEDICINAL

Yes. Contains psilocybin and psilocin, which are in phase II clinical trials to treat depression. Many more clinical trials using psilocybin and psilocin are underway to help treat end-of-life anxiety, post-traumatic stress disorder, anorexia, alcoholism, smoking addiction and a host of other mental health issues.

PSYCHOACTIVE

Yes. Principal active compounds are psilocybin and psilocin, with an onset of varying psychedelic effects occurring within 20–60 minutes.

ENVIRONMENTAL REMEDIATION

No.

SPORING BODY CHARACTERISTICS

CAP

1.5–10 cm wide
Bell-shaped, convex or flat
White with brown centre, bruises blue
May have small white spots

GILLS

Purple-brown
Close
Attached or free from stipe

STIPE

5–15 cm tall
0.5–2 cm thick
White to yellow-brown, bruises blue
Smooth to silky
Thin upper ring

SPORES

Purple-brown to black
Oval

FIELD DESCRIPTION

HABITAT

Grows in cow dung, and occasionally horse and elephant dung.

DISTRIBUTION RANGE

Truly global. Found throughout the tropical and subtropical climate of Southeast Asia, India, Australia and the Americas.

SEASON

Summer and autumn.

PSILOCYBE CYANESCENS

COMMON NAMES

Wavy cap, cyan

FAMILY	Hymenogastraceae
GENUS	Psilocybe
SPECIES EPITHET	cyanescens

When mature, the cap of *Psilocybe cyanescens* lifts up around the edges and develops a signature wave. The surface of the caramel-brown cap is smooth and moist to touch, thanks to a gelatinous film that can be peeled off. It has grown prolifically around the world, and is sought after for its strong psilocybin content. But be careful of its extremely deadly look-alike *Galerina marginata,* which grows in similar habitats and has superficial similarities.

HISTORY AND CULTURE

There is currently no widely accepted understanding of where *P. cyanescens* originated. It seems to prefer artificial areas, growing abundantly on mulch beds and wood chips. One theory is that its spores hitchhiked into a wood chip supply centre and colonised the mulch that is distributed at a commercial scale to control weeds. This explains the frequent massive harvests of *P. cyanescens* in public places.

PROPERTIES

EDIBLE

Yes, but it is psychoactive. Not for the faint-hearted.

MEDICINAL

Yes. Contains psilocybin and psilocin, which are in phase II clinical trials to treat depression.

PSYCHOACTIVE

Yes. The sporing bodies of varieties growing in North America are considered to be among the most potent psychedelic mushrooms. Principal active compounds are psilocybin and psilocin, with an onset of varying psychedelic effects occurring within 20–60 minutes.

ENVIRONMENTAL REMEDIATION

No.

SPORING BODY CHARACTERISTICS

CAP

1.5–4 cm wide
Wavy or upturned at edges
Brown to yellow-brown, bruises blue
Sticky when moist

GILLS

Brown
Close
Attached to stipe

STIPE

2–8 cm tall
0.2–1 cm thick
Off-white, bruises blue
Smooth
May have upper ring

SPORES

Purple-brown to black
Oval

FIELD DESCRIPTION

HABITAT

Grows in colossal clusters, often by the hundreds or even thousands, on wood-based substrates such as mulched plant beds, wood chips and sawdust.

DISTRIBUTION RANGE

Truly global. Can be found throughout North America, Europe, Australia, New Zealand, Iran, northern Africa and Asia.

SEASON

Autumn and winter.

PSILOCYBE MEXICANA

COMMON NAMES

Mexican mushroom, *teonanácatl* ('flesh of the gods' in Nahuatl), *zize* ('little birds' in Mazatecan)

FAMILY	*Hymenogastraceae*
GENUS	*Psilocybe*
SPECIES EPITHET	*mexicana*

Psilocybe mexicana is a small, wispy mushroom that grows to a height of 12 centimetres. It has a bell-shaped, straw-coloured cap. It is legal in the Netherlands through a loophole that allows its sclerotia to be sold as 'magic truffles'. The sclerotia form as hardened masses of mycelium in response to harsh conditions such as drought or lack of nutrients. Depending on the substrate, sclerotia contains roughly half the psilocybin and psilocin content of dried mushrooms. Sclerotia tastes similar to pungent psilocybin mushrooms, but has the distinct texture of walnuts.

HISTORY AND CULTURE

P. mexicana brought psychedelia to the modern Western world. It was the mushroom María Sabina used in her *velada* when the first Westerners, Robert Gordon Wasson and Allan Richardson, joined her healing ceremony in Mexico in 1955. French mycologist Roger Heim joined Wasson on his next visit to determine how to cultivate *P. mexicana* in the laboratory. They also sent samples to the creator of LSD, Albert Hofmann. He successfully isolated and synthesised the active compounds, naming them 'psilocybin' and 'psilocin'.

PROPERTIES

EDIBLE

Yes, but it is psychoactive. Not safe for work.

MEDICINAL

Yes. Contains psilocybin and psilocin, which are in phase II clinical trials to treat depression.

PSYCHOACTIVE

Yes. Principal active compounds are psilocybin and psilocin, with an onset of varying psychedelic effects occurring within 20–60 minutes.

ENVIRONMENTAL REMEDIATION

No.

SPORING BODY CHARACTERISTICS

CAP

0.5–2 cm wide
Bell-shaped or conical
Straw, brown or red-brown, bruises blue
Smooth, slightly translucent, jagged at edges

GILLS

Grey to purple-brown
Distant
Attached to stipe

STIPE

4–10 cm tall
1–3 mm thick
Straw, brown or red-brown, bruises blue
Smooth or silky

SPORES

Dark purple-brown
Oval or leaf-shaped

FIELD DESCRIPTION

HABITAT

Grows along roadsides, trails and in grassy areas near forests, although it prefers meadows that are rich in manure.

DISTRIBUTION RANGE

Native to North and Central America.

SEASON

Spring, summer and autumn.

PSILOCYBE SEMILANCEATA

COMMON NAME

Liberty cap

FAMILY	*Hymenogastraceae*
GENUS	*Psilocybe*
SPECIES EPITHET	*semilanceata*

Widely known as 'liberty cap', *Psilocybe semilanceata* was the first European species confirmed to contain psilocybin in the 1960s. *Semilanceata* means 'spear-shaped' in Latin, referring to its pointed, conical cap perched on a long thin stipe. It looks small and fragile even among blades of grass, but it is one of the most potent psychedelic mushrooms in the world. It can also form sclerotia, the dormant form of the fungus, to weather natural disasters. The high concentration of psilocybin makes the experience extremely visual and it can last longer than that induced by other species.

HISTORY AND CULTURE

P. semilanceata is an iconic species in Europe. Its common name stems from a 4th-century tradition in the Roman Empire, when a 'liberty cap' made of soft felt was given to freed slaves to symbolise their new status in society. Liberty caps were political symbols in the French and American revolutions in the 18th century, and they were placed on a pole to symbolise rebellion. Fittingly, *P. semilanceata* is also the symbol of an inner revolution today.

PROPERTIES

EDIBLE

Yes, but it is psychoactive. Tastes trippy.

MEDICINAL

Yes. Contains psilocybin and psilocin, which are in phase II clinical trials to treat depression.

PSYCHOACTIVE

Yes. Principal active compounds are psilocybin and psilocin, with an onset of varying psychedelic effects occurring within 20–60 minutes.

ENVIRONMENTAL REMEDIATION

No.

SPORING BODY CHARACTERISTICS

CAP

0.5–2.5 cm wide
Bell-shaped or conical, pointed at centre
Brown to tan, bruises blue
Radial grooves, sticky when moist

GILLS

Grey to purple-black
Close or crowded
Attached to stipe

STIPE

4–12 cm tall
1–3 mm thick
Brown to tan, bruises blue
Smooth and flimsy

SPORES

Dark purple-brown
Oval

FIELD DESCRIPTION

HABITAT

Grows individually or in groups in well-fertilised grasslands, but not directly on dung. A saprophyte that feeds off decaying grass roots.

DISTRIBUTION RANGE

Widespread in temperate areas around the world, particularly in the North America and the UK.

SEASON

Summer and autumn.

Fungi can transform a range of organic waste into high-performing materials of almost any shape, strength and density. Production of mushroom materials uses less water, energy and land than traditional materials, and there is no waste in this circular process – all the materials can be composted.

FUNGI CAN SAVE OUR WORLD

It's no secret that when it comes to the natural world, we take far more than we give back. Far from our humble beginnings living in harmony with the land and the ocean, we have now harnessed the power of the human collective to organise ourselves into magnificent cities and cultures. While our ancestors were striking flint to start a fire, we now can create nuclear fusion, the very process that powers our sun and the stars. By most measures, we have performed unimaginably well. Yet from nature's perspective, we have been swept up by the impressive velocity of this advancement and outpaced the steady rhythm of her dance. We have forgotten that nature is not a part of our world, but that we are a part of nature.

You probably don't walk around appreciating the impossibility of life in the vast expanse of our universe. But our temporary lease here on Earth is just that, ephemeral. Our modern civilisation, steeped in culture, technology and beliefs, represents a mere 0.00002 per cent of Earth's 4.5 billion-year history. Our fleeting existence should be in reverence to the processes our planet has been undergoing for billions of years – processes that will continue long after we are gone. Instead, we are in competition with natural systems and outcompeting them.

At the First World Climate Conference in 1979, scientists from fifty countries acknowledged that the alarming trends of climate change made it urgently necessary to act.[1] Since then, alarms have been sounded year after year by scientists, activists and citizens. We are in a self-inflicted climate emergency. We only have one Earth and, unless we find another one to go to, we need to seek solutions in our own backyard.

Earth is a closed-loop system. All physical matter exists in fixed and limited amounts. What we have is what we've got – forever. For life on Earth to continue, matter must be recycled over and over again. Nature is circular – when a plant dies, the nutrients are fed back into the soil for the rest of the ecosystem. Our current supply chain and consumption habits are linear. The end of the 'take-make-use-dispose' model is landfill.

There is no doubt that the industries of the future will choose to implement design principles that underpin a circular economy. It's happening now. These principles include keeping materials in the use cycle, designing out waste and partnering with living systems to minimise environmental pollution.

Luckily for us, there's a natural kingdom with a billion years of experience ready and waiting to share. But can we slow down enough to learn the lessons it has to teach us? Working hard in the shadows, fungi are critical to the natural processes we take for granted. Pioneers in the field of science, biotechnology and business are recognising their power to help solve our problems. Astounding applications of fungi – from removing toxins and pollutants in the environment (mycorestoration) to transforming waste into useful materials and products (mycodesign and mycofabrication) – are possible. A global movement is underway to take on those learnings and save our world.

We only have one Earth and, unless we find another one to go to, we need to seek solutions in our own backyard.

5.1
MYCORESTORATION

Fungi have a range of natural abilities that we can use to heal damaged habitats. This is known as mycorestoration. Scientists can leverage fungi's ability to decompose, and engineer it to break down pollutants in the environment, particularly xenobiotics. Xenobiotics are chemical substances that have been introduced by humans and do not exist in nature, such as pesticides, cosmetics, industrial chemicals and drugs.

The discovery of white rot fungi has propelled the study of mycorestoration. White rot fungi is a group of saprophytic fungi that has the unique ability to break down xenobiotics. Mycorrhizal and parasitic fungi also play a role in mycorestoration, as they can accumulate and degrade toxic metals.

Harbhajan Singh, an environmental engineer, published the first comprehensive book on this topic in 2006 called *Mycoremediation: Fungal Bioremediation*. Paul Stamets, a famous mycologist and entrepreneur, is also a leading voice with his 2005 book *Mycelium Running: How Mushrooms Can Help Save the World*. Despite the growing interest, mycorestoration is still an infant science and remains largely experimental.

Transferring these discoveries from the laboratory into the real world is difficult. It takes a multidisciplinary team of chemists, environmentalists, mycologists, botanists, regulatory experts, site managers and more to execute a project on site. There is no silver bullet, no one-size-fits-all approach and it requires constant and consistent work with nature over many seasons and years.

Scaling up these discoveries for commercial viability is even harder. Grants and institutional investments are hard to come by, and without them, academics switch research areas. Pioneering a new field of science is challenging and takes time, money and grit. But extraordinary circumstances need extraordinary solutions – and fungi can provide them.

FUNGI AS A WATER FILTER (MYCOFILTRATION)

Earth is the only known planet within our solar system that has bodies of liquid water on its surface. This clear liquid is one of our most precious resources, but water supplies are limited. Less than 1 per cent of the water on Earth is accessible and fit for consumption, and this is currently shared between households, agriculture and industry. Over 97 per cent of Earth's water is too salty and 2 per cent is freshwater locked away in groundwater, glaciers and ice caps.[2]

In the last 100 years, the world's population grew fourfold as the world's water consumption grew sixfold. The industrial age and modern plumbing have made way for water consumption at rates that were never possible before. This efficiency, coupled with an increase in demand for water, has resulted in global scarcity.

Flushing the toilet or running the washing machine creates wastewater that is not reusable until it is treated. Roughly 80 per cent of the world's wastewater is left untreated and allowed back into our waterways, putting the health of our water ecosystems at risk.[3] Even in developed countries, wastewater is not properly decontaminated due to outdated treatment plants, sewage overflows and ineffective household sewage treatment systems. The source of untreated wastewater is difficult to pinpoint, as it originates from a range of sources, which often include agricultural and stormwater run-off.

A promising example of an affordable and feasible solution is mycofiltration. This process uses fungal mycelium as a biological filter to capture and remove contaminants from water and soil. Depending on the fungal species, mycelium can even eat through and digest pollutants such as pesticides, mercury and petroleum products. If you peer at fungi through a microscope, you'll see that the cells of mycelium are about 0.5–2 microns wide (for comparison, a strand of human hair is 50 microns wide). Mycelium grows as interconnected cells that resemble a netted fabric.

Armed with this knowledge in the 1970s, Paul Stamets imagined that this fabric of interconnected cells could become a biological filter. He tested this hypothesis on his waterfront farm, installing large sacks filled with substrate inoculated with mycelium of the garden giant species (*Stropharia rugosoannulata*) around water basins. The sacks formed a netted barrier to catch contaminants as water passed through. This mycofilter cleansed the water, resulting in a 100-fold drop in coliform levels – the bacteria that is present in the digestive tracts of animals and found in their waste. The mycofilter successfully reduced fecal matter in the water, alleviating the downstream impacts of contaminated water. This finding was later investigated and confirmed by the US Environmental Protection Agency (EPA).

A mycofilter can be as simple as a hessian burlap sack filled with wet straw and wood chips, and inoculated with mycelium. It is inexpensive and simple to set up. Also, the small size of the mycofilter means that it has minimal impact on ecology and can be installed around sites such as farms, urban areas, roads and factories. Having a mycofiltration system in these areas can help decontaminate wastewater before it makes its way back into our waterways.

Mycelium is known for its insatiable hunger for organic matter. Specifically, the oyster mushroom (*Pleurotus ostreatus*) is able to process and neutralise bacteria such as *Escherichia coli* (commonly known as *E. coli*), working with its mycelial membrane to filter out microbial pathogens from contaminated water.

Tradd Cotter, mycologist and owner of the company Mushroom Mountain, runs workshops on setting up mycofiltration systems. 'We're using a cage that looks like a crab pot, that can be refilled with wood chips. It'll last for a year or two. And if the cage stays put, it can be emptied out and refilled with new wood chips.'[4]

Mycofiltration is a young science and commercial applications are scarce, but this has not stopped property owners from experimenting with this fungal capability.

FUNGI AS A FOREST AND SOIL BUILDER (MYCOFORESTRY)

Forests cover one-third of the land on Earth and their diverse habitats are home to 80 per cent of the world's (known) plant and animal species. As for humans, billions of people in rural communities rely on forests for food, shelter, medicine and water. Forests are also a vital player in the effort against climate change as they act as a carbon sink – they absorb, or sequester, large amounts of carbon dioxide and store the carbon in their wood. Old-growth forests are particularly critical, because their roots have extended deep into the soil for centuries and sequester extra carbon out of the atmosphere, helping to manage today's rising temperatures.

Yet deforestation – legal and illegal – continues. Aside from permanent losses of biodiversity, deforestation sets off a domino effect of land degradation impacts, including increased erosion, reduced soil fertility and piles of wood debris. Unfortunately, the standard treatment of this 'waste' is incineration. This releases additional greenhouse gases into the atmosphere and destroys the potential for the nutrients from the wood to be recycled back into the soil. This calls for sustainable forest management practices. One of these is mycoforestry – the use of fungi as a forest and soil builder.

Wood debris in forests can be chipped into smaller pieces, then inoculated with native saprophytic fungi species to speed up the decomposition process. This redirects vital elements and nutrients back into the soil for use by the rest of the forest. Fungi also produce glomalin at the hyphal tips, which is what fungi use to store carbon. It's a sticky substance that binds together soil particles and builds soil architecture. This aerates the soil, helps water and nutrient retention, and regenerates impacted habitats.

Along with forests and oceans, soils also act as an invaluable carbon sink. Research found that mycorrhizal fungi in boreal forest islands in Sweden hold up to 70 per cent of the total carbon stored in soils.[5] This means that trees connected to a mycelial network absorb carbon from the atmosphere and then transfer it into the mycelium for storage.[6] Fungi play a critical role in regulating the global climate.

Mycoforestry was put into action in Colorado by Jeff Ravage, North Fork Watershed coordinator for the Coalition for the Upper South Platte and researcher at Denver Botanic Gardens. In 2016, he and his team of researchers set up two test sites in Denver Mountain Parks.

These sites had been logged and abandoned, leaving 30 centimetres of waste wood spread across the entire forest floor. Ravage's team recruited the help of wood-rotting species *Pleurotus pulmonarius* on one site and *Morchella angusticeps* on another.

Over five years, fungi in the first site consumed the wood chips and created 5 centimetres of topsoil – organic, mineral-rich soil from which seeds germinate – with a few centimetres of partially decomposed organic matter on top. Before this, the ground was just gravel with dust. The second site exhibited a slower rate of seeding, decaying 75 per cent over two years. The team will be treating this site with a second inoculation in 2021 to further investigate the results.

Want to get started on your own mycoforestry project? Watch out for a paper from Ravage, who wants 'to create useful tools and distribute them freely, because we don't have enough time left for somebody to figure out how to make a profit on saving the planet. We're not out to create a patent,' he says. 'How can we patent nature?'[7]

Mycoforestry remains an experimental forestry practice conducted by environmental groups and volunteers. Replenishing the soil in forests, improving soil fertility and increasing forest ecosystem resilience is of both ecological and economic interest. A greater uptake of mycoforestry by forestry management groups, logging companies and council decision-makers will move the science forward and protect the future of forests.

FUNGI AS A TOXIN NEUTRALISER (MYCOREMEDIATION)

We may not see it, taste it or feel it, but we are entangled in an array of environmental toxins. Microplastics in waterways, nanoparticles in the air and noxious chemicals in soils were all introduced by human activity and have become accepted as invisible causes of illness and death. Contaminants and pollutants are abundant in our air, water and soil, all around the world.

Traditional methods of remediation, such as disposal into hazardous waste facilities, incineration and chemical treatments, are expensive, energy-intensive and only move the contamination to someone else's backyard. We urgently need to find more permanent solutions to clean up the mess we've made on Earth. Many scientists have turned to mycoremediation, the use of fungi to decontaminate the environment. After all, fungi are nature's decomposers and this strategy has been effective for Earth for over a billion years.

In forests, a major source of nutrients is from fallen trees, released as the trees break down. Their sturdy trunks are reinforced by lignin, a complex material that binds together the building blocks of wood. Only fungi can excrete enzymes powerful enough to decompose lignin. Luckily for us, the bonds in lignin are similar to those in petroleum, pesticides, plastics, dyes and a range of other toxins, which means mycelium can disassemble the hydrocarbons present in a wide spectrum of toxins. In particular, saprophytic fungi varieties called white rotters, such as oyster (*Pleurotus ostreatus*) and turkey tail (*Trametes versicolor*) mushrooms, are relatively easy to grow and love molecular decomposition.

In 2016, Fungaia Farm, a mushroom farm in California, used oyster mushroom spawn to remediate gallons of diesel fuel that had spilled from a storage tank. They removed the contaminated soil and placed it between layers of fresh straw and burlap that were inoculated with oyster mushrooms. The mycelium got to work. As it fed on the petroleum, hyphae threaded throughout the crevices of the oil-laden soil. Later testing showed that all contamination was reduced to a non-toxic level and some soil was even oil-free, allowing the land to be reclaimed for landscaping.

Levon Durr, the owner of Fungaia Farm, has noted that the project wasn't without its mistakes and he has since published a report to aid grassroots practitioners.[8] Another diesel spill was discovered in 2020 and the Fungaia Farm team hopes to detoxify the soil using mushroom spawn once again. But convincing landowners to try mycoremediation is challenging. 'It can cost US$15,000 for one remediation treatment on site and it quickly adds up because it's a biological process and may need multiple treatments over the course of a year to get the soil to a non-toxic level,' says Durr, 'compared to paying US$45,000 once-off to dig up the contaminated soil and haul it away.'[9]

Conditions are also difficult to control in these outdoor projects. It may be cold and rainy one month, but dry and hot the next. If temperatures are too high, the piles of soil and burlap can turn into a compost heap and kill the mycelium. Controlling for a myriad of variables on the field takes patience and resilience. Fungaia Farm continues to educate and produce mushroom spawn to cultivators for food production and mycoremediation projects.

Mycocycle, founded by Joanne Rodriguez with Peter McCoy as chief science officer, is pioneering a new industry: using fungi to divert waste from landfills. They are remediating waste from roofing, asphalt and chemical manufacturing industries. As the mycelium consumes waste and binds it together, Mycocycle creates new materials from the process. There is strong interest from manufacturers in these industries for a cost-effective and sustainable waste-treatment solution. The challenge Rodriguez faces to scale up mycoremediation 'is the lack of interdisciplinary backgrounds to move these discoveries out of the lab and into real world treatments'.[10] To combat this, Mycocycle launched an equity crowdfunding campaign in 2020 encouraging people to join the cause and accelerate change. McCoy is also the founder of MYCOLOGOS, an online education platform for all things fungi.

FUNGI AS A PLASTIC DEGRADER (MYCOREMEDIATION)

Despite only being around for the last seventy years, plastic is literally everywhere. Plastic has revolutionised our world: it is tough yet flexible, durable yet easy and cheap to produce. The manufacture of the wide-ranging plastics family begins by drilling for crude oil. The oil is then heated to form chains of carbon called plastic polymers.

In 2017, a global analysis of all plastics ever made showed that 6.6 billion tonnes, or 79 per cent, has accumulated as waste in the environment, 12 per cent has been burned and only 9 per cent has been recycled.[11] While they're undeniably durable, certain plastics, such as plastic bottles and cutlery, have an extremely short useful life. Half of all plastic manufactured becomes trash in less than a year.[12]

With plastic production only projected to increase, research into effective solutions to decompose plastic is critical to alleviating the pollution in our waterways, oceans and cities. Waste is typically burned, which is energy-intensive and emits harmful pollutants into the environment. So far, the only green way to decompose plastic is through photodegradation, which uses UV rays from sunlight to break apart the plastic molecules. Unfortunately, since waste in landfills rarely sees the light of day, it can take up to a thousand years to decompose. These materials do not exist in nature, so nature's organisms have not evolved the ability to break them down effectively, if at all. That is, until recent discoveries of particular fungal species that show an appetite for plastic.

In 2017, scientists from the Kunming Institute of Botany in China discovered a microfungi, *Aspergillus tubingensis*, in a Pakistani rubbish tip that can break down polyurethane.[13] The fungus secretes enzymes that break apart the bonds between the plastic molecules, allowing it to ingest the plastic as food. The fungus also uses the physical strength of its mycelia to help break apart the molecules, making the decomposition process as short as a few weeks. The team, led by Dr Sehroon Khan, has since found fifty other strains of fungi that can break down other types of plastic and is working on finding the optimal environment for these fungi to thrive. The only catch is that funding is limited. Finding a fungus that can break down plastic in a petri dish is vastly different to deploying it at a commercial level in landfills.

Another solution is to design plastics that can be biodegraded using bacteria and fungi. This reduces not only plastic pollution but also our society's reliance on crude oil to synthesise plastics. Crude oil forms when dead organisms are buried and subjected to intense heat and pressure over millions of years – hence the term 'fossil fuels'. Inevitably our brief love affair with plastics will end, just not soon enough.

FUNGI AS A PEST CONTROLLER (MYCOPESTICIDES)

A pest is any organism that causes destruction and spreads disease. The Food and Agriculture Organisation of the United Nations estimates that 20 to 40 per cent of global crop production is lost to pests each year, costing the global economy around US$220 billion.[14] Common prevalent pests include insects, weeds and microbial pathogens, which attack food crops, livestock and building structures.

In 2019, the global pesticides industry was worth US$84.5 billion, largely made up of synthetic chemical products that help farmers control or kill these pests.[15] These synthetic solutions are effective, but they have the same problem as antibiotics – overuse encourages resistance, increasing a pest's tolerance to pesticide exposure. This cycle requires progressively more chemicals to reach the same level of effectiveness, releasing more toxins into the atmosphere and poisoning ecosystems.

Picture your local farmer spraying a chemical mix on your favourite eggplant crop to keep away the notorious potato beetle. These chemicals often run off into nearby waterways, harming fish and animals and making otherwise drinkable water unsanitary. And when you eat the eggplant, the pesticides end up in your body, too. Ongoing studies also show that pesticide exposure is linked to birth defects, cancer and ADHD.[16]

We urgently need pesticides that are less harmful to our environment. Biopesticides are defined by the EPA as 'pesticides derived from natural materials such as animals, plants, bacteria and certain minerals'.[17] One promising solution is mycopesticides, a biopesticide using the green mould fungus *Metarhizium anisopliae* as the active ingredient. This fungus is an entomopathogenic species – a parasitic fungus that feeds on insects and other pests. As insects come into contact with the entomopathogenic fungi, the fungal spores attach themselves onto the insect, grow inside their body and feed on their organs until they are paralysed or dead.

Fungi in biopesticides can be trained through selective breeding to tell the difference between 'good' and 'bad' insects so that they don't disrupt butterflies, pollinators and other beneficial insects. Compared to chemical pesticides – 95 per cent of which do not hit their target –[18] fungal biopesticides have a lower impact on the environment, are not harmful to humans or other animals, and are also considered safe for organic farming. Researchers are rushing towards commercialising new applications of this fungal innovation. A notable philosophy from leaders in the field is to not wage war against the insect kingdom but to enlist fungal defences when pests threaten people, economies or the environment. The aim is to seek balance, not extinction.

FUNGI AS A FARMER'S FRIEND (MYCOPERMACULTURE)

Our food system faces the challenge of ensuring sufficient production and fair distribution of food while protecting nature and preventing climate change. The consequences of feeding more people while drawing on Earth's finite resources are severe. We have appropriated space to meet the demand for energy, water, agricultural land and raw materials. To combat these impacts, there has been a surge of interest in designing housing, food and social systems with nature in mind – a movement called permaculture.

Permaculture is a departure from traditional agricultural practices that was started in the 1970s by Australians Bill Mollison and David Holmgren. It is underpinned by a set of planning and design principles that are inspired by nature. The basics of permaculture involve restoring soil, conserving water, reusing waste and planting year-long food crops. This builds resilient, self-sufficient systems and creates abundance for the grower. It is a form of activism that decouples the grower from the existing industrial food system and its extensive environmental impacts. Permaculture creates self-sufficiency, regeneration and interconnectedness that is not limited to food and can be applied to all aspects of life. Mollison and Holmgren hoped that it would become the 'permanent culture' of how we think and live.

The influence of permaculture is growing and runs deep through urban food gardening and sustainable farming circles. You may know someone who has set up rain barrels, solar panels, composting bins, worm trays or their own chicken pen. Many people like to cultivate mushrooms and microgreens and even experiment with fermentation. The prevalence of fungi in permaculture systems spawned the term 'mycopermaculture'. As nature's recyclers, fungi close the loop and ensure harmony in all living environments.

Imagine a mycopermaculture system in your garden, one which recycles garden waste into nutritious food for fungi. As the mycelium consumes your garden waste, it creates a launchpad for the fungi sporing bodies to develop, creating an abundance of potential edible and medicinal mushrooms for you and your family, and fodder for livestock roaming your garden (animals love mushrooms). Not only that, the by-products can be put into your garden's soil to enhance its nutrients and microflora.

Fungi have proven their importance in the rise of today's permaculture movement. Mycorrhizal fungi increase the resilience of plants and the rate of nutrient cycling in soils, while saprophytic fungi accelerate the decomposition of organic matter.

Small and slow systems are easier to maintain than big ones. With some patience, a keen beginner's mind and the right resources, even city-living, first-time gardeners can make a big difference. The permaculture revolution is possible. It can liberate us from our society's conditioning of over-consumption and waste. Actions at an individual level are a vote for the type of person and society we wish to become.

HOPE FOR THE FUTURE

Perhaps most importantly, fungi provide us with the hope that we can not only survive, but thrive, despite the challenges that we face on our planet. Even though mycorestoration is a new science, small-scale fungal restoration led by grassroots practitioners is occurring around the world. With each strand of hypha, we take another step towards healing our planet. The challenge for remediators around the world is to achieve restoration at scale with cost efficiency and minimal additional disturbance to the biosphere.

As nature's recyclers, fungi close the loop and ensure harmony in all living environments.

HOW TO CULTIVATE MUSHROOMS WITH COFFEE GROUNDS

Sometimes all we need is a perspective shift. Waste doesn't exist in nature, everything is re-used. Similarly, household waste can be repurposed into meaningful inputs to create your own food. You can do this with materials readily available at home or purchased for a low price.

Using the principles of mycopermaculture, we can recycle a common household waste item – used coffee grounds – to cultivate mushrooms in less than eight weeks. The by-product can also be used as a biofertiliser for your garden, closing the loop of this circular process.

WHAT YOU NEED

WHAT	HOW MUCH	WHY	NOTES
Used coffee grounds	A steady supply – however much you have	Used coffee grounds are nutrient-rich and have already been sterilised and hydrated in the coffee-brewing process.	Make sure the coffee grounds are cool so they don't kill the mycelium. Fresh coffee grounds are best. If you can't use them within 24 hours, refrigerate or freeze to prevent mould from forming.
Used coffee filters (optional)	Whatever you have available	Mushrooms can feed on the cellulose and carbon in paper and cardboard.	Soak the filters in water for 20 minutes before use.
Glass jar	One 1-litre or 2-litre jar with a metal lid	The jar will hold the coffee grounds and mycelium spawn.	You can use a preserving or mason jar.
Micropore tape	10 cm	This tape covers the holes in the lid and must be thin enough for the mushrooms to grow through.	You can find this tape at your local chemist or homeware store.
Oyster mushroom grain spawn	50 grams	Mushroom grain spawn looks a bit like little brown peanuts. It contains mycelium and will begin the process of colonising your coffee grounds. These are like the seedlings from which the mushrooms grow.	Search online or contact your local mushroom producer. If you're not using the grain spawn right away, it can be stored in the fridge.
Gloves (any material)	A few pairs	Gloves will keep the equipment sterile and stop your hands from drying out.	Sterilise your gloves every time you touch any equipment.
Isopropyl alcohol (70%)	One small bottle	This is used to sterilise your equipment and gloves between each step.	Everything you touch must be sterilised so that competing microorganisms do not hijack the cultivation process.

WHAT TO DO

Preparing

1 Sterilise your gloves using isopropyl alcohol between each process. This is important to keep out harmful microorganisms.

2 Cut two circular holes into your lid (each about 2 cm wide), evenly distributed across the centre line of your lid.

3 Sterilise the jar and lid in boiling water for at least 10 minutes, ensuring it is submerged. Dry thoroughly.

4 Cover the holes on the lid with two pieces of micropore tape, just enough to cover the holes.

Inoculating and incubating

5 Break up your mushroom spawn if it is clumpy and mix it with your used coffee grounds. Use this mixture to fill the jar up 2–3 cm to begin with. You don't need too much mushroom spawn – a ratio of 1:10 mushroom spawn to coffee grounds is enough. But use more spawn if you can, as it fights against competing microorganisms. You can also sprinkle additional spawn on the top layer of the mixtures and the inner sides of the jar, so that the mycelium grows inwards and traps any contaminants.

6 Cover the jar with the lid and store it in a dark room or cabinet for 2–4 days to allow the mycelium to colonise the coffee grounds. Keep the temperature between 24–27 °C. Too high a temperature might kill the mycelium and allow the growth of contaminants; too low might slow down the colonisation process.

7 Check the jar every day to see if the mycelium has increased its colonisation. The coffee grounds will slowly get enveloped by white, cottony threads. It's exciting to watch it grow from day to day.

8 When the coffee grounds have been fully colonised, add another 2 cm of coffee grounds on top. Close the jar, return it to the dark area and wait for the mycelium to colonise it again. Repeat this process until your coffee grounds fill the jar and are completely covered with mycelium. This may take a few weeks. Note: Refilling the jar in increments ensures you use up your coffee grounds as you go. There's no waste. If you have enough coffee grounds to fill up the entire jar in one go, simply mix in the appropriate ratio of spawn and follow the same process.

Growing mushrooms

9 Once the grounds are fully colonised, place the jar in a humid area with temperatures between 17–20 °C. It is important that this area is sunny, but do not place the jar in direct sunlight.

10 Check the jar daily. After a few days, you should see tiny pinheads forming. Over the next week, the mushrooms should double in size every day. After 10–12 days, the mushrooms should make their way through the taped holes. If the mushrooms are struggling to push through, remove the tape.

11 The mushrooms are ready to be harvested when they form a cluster. The caps will also begin to turn upwards. Twist the cluster to pull off your freshly grown mushrooms.

Repeating

12 This first batch of coffee grounds can yield a second or third flush of mushrooms. Once you've harvested your first batch, return the jar to the same sunny area. If you don't see a second flush, fill the jar with water and drain it after 12 hours to shock the mycelium into growing. Repeat this step to trigger a third flush.

13 After the third flush, the mycelium will probably not produce any more mushrooms. Remove 80 per cent of the mycelium coffee grounds. Don't throw it away – break it up and use it as a biofertiliser for your home garden.

14 To grow a new batch, continue from Step 8. No new spawn is necessary. This will start a new process of colonising fresh coffee grounds. Don't forget to put new micropore tape over the holes in the lid.

5.2

MYCODESIGN AND MYCOFABRICATION

The products used to maintain the developed world's standard of living, from home decor to electronics, are all created from Earth's raw materials. These materials undergo industrial processes that incur massive hidden costs that escape the collective awareness. For example, the most commonly used structural materials – concrete, steel and aluminium – contribute more than 22 per cent of global carbon dioxide emissions. As for the fast fashion world of $10 t-shirts, producing 1 kilogram of cotton in India – where most cotton is grown – consumes 22,500 litres of water, with no opportunity for reuse.[19]

Instead, the future can be grown. Designers interested in the future of materials are innovators. They are taking inspiration from nature's design systems and manifesting a reality based on a circular economy of materials. Rather than using up finite resources such as fossil fuels, land and animals to meet our needs for materials, we can grow them through partnerships with living organisms. Inevitably, industrial manufacturing will be replaced with biological manufacturing. Materials will be grown, fabricated, used and, at the end of their lifecycle, biodegraded or reused – closing the loop of the circular economy.

The ever-adaptable mycelium has become a core building block for the new biological materials revolution, also referred to as mycodesign and mycofabrication. Using mycelium, fungi can transform a range of organic and inorganic waste into high-performing materials of almost any shape, strength and density. The process is energy- and resource-efficient. Mycodesign and mycofabrication form a new class of materials and products that meet a far superior level of sustainability compared to traditional products and their manufacturing methods.

The processes to create mycelium materials are complex, but here is a simplified version. To begin, substrates can be prepared using different types of organic waste, such as corn cobs, egg cartons or agricultural waste. The waste is inoculated with spores of fungal strains that have been selected for their growth speed and compatibility with the substrate. From there the mycelium grows quickly, weaving and intertwining itself through the substrate. As the mycelium grows, it acts as a natural glue that holds the substrate together, self-assembling into the shape of whatever container it sits in. The starting materials convert into advanced biomaterials in less than two weeks. The new material is then removed from the container and dried at the right temperature and humidity.

LEFT
These images show three stages of mycelial growth in hemp mulch:

Top layer: The mulch is inoculated with spores.

Middle layer: Mycelium consumes the mulch over twelve weeks.

Bottom layer: Mycelium completes colonisation and assumes the dimensions of the container.

This process is one way mycomaterials can take the form of flexible sheets, garment materials and packaging. Depending on the type of substrate, fungal species used and the custom grow process, the properties of the material can change to suit different requirements and performance. For example, using heat pressing will strengthen and harden the composite material, causing the foamy structure to harden like wood.

At the end of the material's useful life, it is 100 per cent biodegradable and compostable. This regenerative design process creates systems that are not only efficient, but also waste-free.

Composite mycelium materials are a blend of mycelium reinforced with another material, such as cotton, wood or even metal. This level of customisation gives the grower more control over the manufacturing process and can produce materials that may rival and replace traditional materials, such as plastics and timber, for specific industries.[20]

MYCOARCHITECTURE

Architecture is both a science and an art, involving design, planning and construction. As we marvel over structures, such as the Metropol Parasol in Spain designed by German architect Jürgen Hermann Mayer, we forget the vast amounts of time, energy and resources they took to create. The Metropol Parasol consists of six parasols in the form of giant mushrooms and it is one of the largest wooden structures in the world. If we hope to maintain today's quality of life, we will need to challenge the current ways we manufacture, use and regenerate materials more than ever.

Engineered wood is one of the most common building materials. It is an all-rounder that is versatile and available in a variety of strengths, thicknesses, sizes and durability. But the production of engineered wood presents an array of problems. To obtain wood, natural forests are logged. Deforestation contributes 15 per cent of global greenhouse gas emissions each year.[21] Engineered wood only consists of 85 per cent natural wood. The remaining 15 per cent is made up of toxic resins derived from fossil fuels. These resins are binding agents and they release a gas called formaldehyde, a known human carcinogen. Another building material culprit is polyurethane. This is plastic foam made from crude oil that is used in wall insulation. Three million tonnes of it is produced in the US each year, of which 80 per cent ends up in landfills and 20 per cent ends up in waterways.

What if architecture could grow, decay and then grow again? Researchers in the mycofabrication field have transformed mycelium into organic materials for use in different construction applications. Mycelium can be engineered to match the strength of alternative materials and has properties that make it non-toxic and resistant to fire, mould and water.

Based in Italy, Mogu is a leading innovator of mycelium-based technologies, materials and products. They partner with mycelium to create a large variety of composite materials for use in interior architecture. The first of their kind in the flooring industry, Mogu floor tiles are fully certified, sustainable alternatives to traditional solutions manufactured using fossil fuels.

Mogu created these floor tiles by joining a dense mycelium-composite core with a top layer coating of a 90 per cent bio-based resin. This is a big step forward compared to the 20 to 30 per cent bio-contents found in the market today. The mycelium-composite core is made of low-value organic residue such as cotton and hemp inoculated with specific fungal strains. The mycelium feeds on the organic matter and binds it into a densely compressed board. The top layer coating incorporates fillers such as corn crops, rice stray, hazelnuts, oyster shells and spent coffee grounds. Maurizio Montalti, co-founder of Mogu, says 'the mycelium-composite core board, constituting most of the volume of a single Mogu floor tile, surpasses the technical performances offered by traditional engineered woods,' and emphasises, 'our products are not to be looked at as replacements for materials traditionally used in architecture, but as responsible alternatives for clients who value a circular approach'.[22] Mogu proves that it is possible to be responsible without compromising technical performance and luxurious design aesthetics, while still minimising one's environmental footprint.

Mogu's ethos is to develop and design products in line with the principles of a circular economy. Mogu strives to use and transform raw materials that come as cast-offs from other industrial processes and value chains. Montalti prefers to not use the word 'waste' – he says 'waste is a man-made notion that should be obsolete, because waste does not exist at all in nature.' He prefers instead to describe them as 'low-value residual matters' that can be upcycled.[23]

Mogu is also designing a process for their floor tiles to be reclaimed and recycled at the end of the product's life. The mycelium-composite core will be separated from the bio-based resin, then ground down and upcycled to create new floor tiles. Accordingly, Mogu has earned a Blue Angel ecolabel certification for their floor tiles, which confirms that they are low in emissions and pollutants and have no adverse impact on the environment.

UK biomanufacturing company Biohm has developed a fungal insulation panel to solve the polyurethane issue. Founder Ehab Sayed explains that they start with a waste stream and feed it to the mycelium. Once the mycelium creates the desired material, they use heat or dehydration to cure it. Curing breaks down the cell walls of the fungus, which deactivates the living organism and makes it ready for installation. This natural process does not include any chemicals or additives and takes thirty to fifty days from waste processing to deliver the product, depending on the aggressiveness of the species. Sayed calls this 'matchmaking' – marrying a perfect substrate with a perfect fungal strain creates a perfect material for a specific application.

Described as 'mushroom insulation', Biohm's fungal insulation panels have outperformed alternatives on the market across a range of properties. Early tests show that fungal insulation can keep you warmer, outperforming alternatives by two to three times based on its thermal conductivity rating. Its fire performance is superior, as there are no synthetic chemicals, and, on average, it halves the spread of fire in a building. Its volatile organic compounds also achieved an A+ rating, which means it releases zero airborne toxins and increases indoor air quality, making it one of the healthiest insulation materials on the market.

Some people might wonder if the durability and life span of a natural product would be shorter than synthetics, but Sayed says that Biohm's fungal insulation outperforms plastic products over time. 'Mycelium only degrades when it's in contact with the right microbiome and much like any other natural or synthetic material, when exposed to repetitive extreme heat and humidity. So if it's in a dry, cool or warm state, it will remain performing as expected for its application. I think it's a misconception that often stops bio-based materials from advancing.'[24]

Sayed's philosophy is to allow nature to lead innovation and make sure everything Biohm touches has a positive or regenerative impact. This is echoed throughout their processes, from embedding the principles of the circular economy to working with local communities to scale their production facilities. These community facilities are a one-stop shop for processing waste and producing mycelium insulation panels – they are even offered a 50 per cent profit share on every product they sell.

Biohm's fungal insulation panels are currently priced on par with premium synthetics but they are hoping to be competitive at the lower end of the market as they scale up their operation. They want their product to be accessible for people in social housing, whose homes tend to be less well built or maintained and who therefore need it most. Sayed offers a gentle reminder about the intricacies of working with mycelium. It releases carbon dioxide as it grows, so it's important to build in processes that can capture the carbon produced.

Mycoarchitecture does not stop at the boundaries of our planet. When humans become a multi-planetary species, new architectural design and production principles will need to be created and applied in space. In 2017, the European Space Agency, with Utrecht University and Officina Corpuscoli (a design-research studio run by Maurizio Montalti), conducted a feasibility research project to explore the opportunities for growing fungal structures on new planetary surfaces, such as Mars and the Moon, to create new habitats.[25] Results show that the growth of biocomposite materials involving the fungal species *Schizophyllum commune* was successful in extreme conditions of microgravity, macrogravity, temperature and radiation, such as cosmic gamma radiation. This early research shows that fungi can be successfully leveraged to create building materials, but much more multidisciplinary work needs to be done before mycelium materials can reliably surpass the performance of conventional building materials, such as concrete.

A paradigm shift is occurring as more and more companies participate in the circular economy. The definition of waste is being rewritten as we work with fungi to create new bio-based materials. An increasing awareness of these materials and their potential applications allows us to design *with* nature rather than against it. What's more exciting is that fungi is at the centre of this. The pioneers believe that we haven't even scratched the surface of its potential applications.

MYCOTEXTILES

The glamour of the fashion world makes it easy for the industry to distance itself from the reality of who makes the clothes, what resources are used and where the clothes go after each collection. The fashion industry is one of the biggest polluters in the world after the petroleum industry. These issues are growing as fast as our consumption habits – habits that need to change urgently.

In a true circular economy, all inputs, from sourcing the raw materials to end-of-use treatment, are infinitely recyclable. Biofabrication, the burgeoning blend of biology and fabrication, is paving the way for a future where materials can be sourced from partnerships with living organisms. We can literally grow the resources we need for the fashion industry using fungi, without compromising on quality or durability. With materials made from fungi, consumers won't even need to settle for vegan leather (which is typically made from fossil fuels).

Debuting at New York Fashion Week 2020, Reishi by MycoWorks holds promise for luxury fashion. Reishi is a new class of mycelium materials that performs, feels and even smells like soft leather. MycoWorks's patented Fine Mycelium technology engineers mycelium to form interlocking cellular structures, which creates a strong sheet of mycelium material in just a few weeks. It is then tanned just like real leather in a tannery. Reishi can be exposed to heat, glues and dyes to produce materials in a range of colours, patterns and tactile finishes previously impossible with animal hides.

French luxury fashion house Hermès has paired with MycoWorks to create an entirely new plant-based material called Sylvania. The result of three years' collaborative work between the companies, Sylvania is made with Fine Mycelium using MycoWorks's patented process, then tanned in Hermès workshops. The fashion industry and consumers alike have been waiting for a sustainable leather option, and this collaboration pairs expertise in tanning the highest quality materials with the tools of biotechnology. Sylvania debuts in the Autumn–Winter 2021 collection, paired with H Plume canvas and Evercalf calfskin in a new version of the Hermès Victoria bag.

In 2016, Bolt Threads also joined the ranks of mycelium-based leather companies, producing Mylo leather by licensing Ecovative's MycoComposite mycofabrication technology. Using dust and organic materials as the nutrient base, the mycelium consumes the substrate in just a few days, creating an interconnected mass. The mass can be transformed into any size, shape or density, producing a leather-like finished material. Bolt Threads developed their global supply chain with a healthy agricultural economy in mind. To that end, they are working with traditional farmers to pivot and innovate their current business to be more sustainable and futureproof, and to create a robust global network of Mylo producers. In 2020, Bolt Threads unveiled a partnership with Kering, Lululemon, Adidas and Stella McCartney to collectively invest in, produce and launch Mylo products into the market in 2021.

It is a busy time for the mycelium leather space. Ecovative announced their re-entry with Forager Hides, which they will produce in their extensive production facilities in the hope of pushing the industry forward. The entire Forager Hides production process takes nine days, compared to two years for animal leather and six months for cotton. As mycelium is alive and self-assembling, it also uses a fraction of the water, land and energy compared with these other textile manufacturing processes. Gone are the scraps from the cutting room floor – mycelium is grown in moulds to exact size and shape specifications. Ecovative has been expanding their operations since 2010 and they are here to stay.

It's an exciting time to join the biotechnology revolution and transition into a circular future. This is still a young field and change doesn't happen overnight. Few companies have cemented partnerships with brands and built new supply chains to bring products to the market. Commercialisation is the second hill to climb after product development. Luckily, the fashion and textiles industry thrives on a quick pace and conscious consumers are waiting for fashion to go fungal. At the end of the day, the consumer makes the rules by choosing what to buy, so ethical fashion starts with all of us.

MYCOPACKAGING

The need for packaging arose when production and consumption took place at different locations. With the increasing rate of urbanisation, packaging is used more than ever. Today, 40 per cent of the plastic produced for packaging is used once, then discarded.[26] In the food industry, the goal is to prolong the shelf life of food to minimise food waste, which is indeed a global issue. But how can plastic, with a useful life of a few days, be used in such quantities when it takes hundreds of years to break down?

It's not just plastic and it's not just the food industry. Packaging is everywhere. The traditional shopkeeper in the marketplace has been replaced with packaging, taking on the functions of protection, distribution, education and marketing. To the consumer, product quality is not only determined by the product itself, but also by the packaging. Ultimately, the purpose of packaging is transient. We need a solution that is easily manufactured, effective during its lifetime and then easily reused or biodegraded. Recycling should be a last resort. Plastic can only be recycled nine times at most and each recycling loop requires energy.

At the crux of all this, the challenges faced by alternative packing materials revolve around four themes: sustainability, strength, design and cost. A clear research and development leader in the mycelium packaging space is Ecovative, whose MycoComposite technology grows mycelium in a mould to create a custom structure called Mushroom Packaging. Co-founder and CEO Eben Bayer describes the action as '5 per cent mycelium pulling together 95 per cent wood chips'.[27] The mycelium acts as a natural glue, forming strong, sturdy materials that can hold products safely and securely in transit.

Up to a threshold of a few hundred thousand units per year, manufacturing mycelium packaging is much faster and more cost-effective than plastic. It uses less energy and produces up to 90 per cent fewer carbon emissions. As for end-of-use, Ecovative's Mushroom

Packaging does not require a composting facility. Instead, it relies on microbes in soil for decomposition. After customers receive their package, they can break up the packaging, throw it in some soil and in thirty days it will be composted. In contrast, typical packaging materials such as styrofoam take centuries to decompose and are non-recyclable.

Companies around the world have licensed Ecovative's technology: Magical Mushroom Company in the UK, Grown, based in Europe and Paradise Packaging Co. in the US, and there are others emerging who plan to service the Asia-Pacific, northern African and southern European regions. With these providers on board, more local businesses will be able to source Mushroom Packaging. Bayer welcomes the cultural zeitgeist in which raw and organic-looking materials are sought after. Mycelium-based products naturally have this finish. This is in contrast with the packaging industry 'seven years ago when the demand was for a packaging material that looked more like styrofoam'.[28]

This eco-conscious boom will aid our move away from fossil fuel reliance and minimise the risk of harmful microplastics entering our soil and seas. As well as the obvious sustainability benefits of a circular production process based on fungi, all the questions about its strength, design and cost have been answered. It's time to replace our current packaging.

NATURE CO-DESIGN

With the triple threats of climate change, resource depletion and waste build-up, we must radically transform how we interact with nature. Fungi are catalysts for change. Their natural design principles and manufacturing capabilities are being engineered to create the upcoming biotechnology revolution.

Using minimal energy, fungi can be engineered to produce high-performing materials with precision. Atom by atom, we can create a new class of sustainable materials, much like a biological 3D printer. This is a part of the 'nature co-design' biotechnology revolution advocated by Massimo Portincaso, a partner at Boston Consulting Group who co-leads the Deep Tech practice. In the paper *Nature Co-Design: A Revolution in the Making*, he refers to nature co-design as the next wave of innovation, 'where biology, material science and nanotechnology meet to leverage nature's design principles and manufacturing capabilities at the atomic level'. This frees us from our reliance on extracting raw materials for energy and products and allows us to use the inherent manufacturing capabilities of living organisms. The paper refers to this biotechnology revolution as the fourth industrial revolution, after coal, electricity and the internet, with the potential to disrupt every industry in every economy, and a total market opportunity of US\$30 trillion over the next thirty years.[29]

We are moving towards a collective fungal consciousness that recognises the wisdom and potential of fungi. The future is bright when we have great minds designing our future in partnership with nature.

5.3

CONSERVATION OF FUNGI

Despite the essential role that fungi play in our ecosystems, it is estimated that 90 per cent of species remain undocumented. With the accelerating impacts of climate change, habitat destruction and pollution, fungal species are disappearing without a trace, unaided by legal protection. According to the *State of the World's Fungi 2018* report by Kew Gardens in London, 'only 56 species of fungi have had their conservation status globally evaluated for the International Union for Conservation of Nature Red List, compared to 25,452 plants and 68,054 animals'.[30]

With 75 per cent of human perceptions based on vision, we commonly neglect the things that are invisible to the naked eye. But with each loss of a fungal species, we lose more of nature's ability to recycle, decompose, strengthen forests, distribute water and nutrients and clean our air. We also lose future discoveries of new medicines, food and materials.

Fungi provide enormous hope for saving our world, so let's give them a real chance. As we see more of fungi embedded into modern society, as packaging for our online orders or replacing leather in our garments, or as the building blocks of eco-friendly architectural projects, perhaps we'll become more aware of the immense kingdom that surrounds and supports us.

By becoming conscious consumers, reconnecting with our forests and furthering our knowledge, we can help save fungi too.

LEFT

Colours and textures of *Trametes versicolor*, a species commonly used in mycofabrication, enclosed in a glass display cube.

ON PAGE 191

Schizophyllum commune has distinct radiating gills. It is one of the most prolific mushrooms on the planet.

Flora, fauna, funga

GIULIANA FURCI

It's rot 'n' mould, not rock 'n' roll!

Energy is not lost, it is transformed. Fungi lead this essential process. Fungi are the organisms that primarily decompose matter to enable recomposition, that chemically hack systems to produce changes that enable the continuation of life. As if that is not enough, they also assist the life of plants and animals through essential symbioses that enable the existence of the ecosystem. They build – and burn – the bridges of life.

However, fungi are not exempt from threats to their wellbeing. They are as affected by climate change, habitat loss and fragmentation, nitrogen enrichment and use of fungicides as plants and animals. Human impacts lead to species loss in Kingdom *Fungi*. It is estimated that we have only identified 10 per cent of all fungal species, so it is highly likely that we are losing species of fungi that we didn't even know existed.

What can we do? Conserving fungal habitats is the key, because fungi are not separate from their symbionts. They exemplify the interconnectedness of life. For the most part, they are specific to their substrates, making every species of plant and animal an ecosystem in their own right. To make sure the pine mushroom thrives in nature, we must make sure the pine tree thrives. To make sure the caterpillar fungus thrives, we must ensure the life cycle of the moth. Fungi demonstrate that we are not independent of others.

Unfortunately, in the eyes of most governments, if there is nothing under threat, there is nothing to protect. To ensure the legal protection of fungi, we have to make the case that they *need* protection. This process has been challenging, because fungi are explicitly excluded from policy frameworks because they are neither flora nor fauna. Rather, they are funga – a term that accounts for the fungal diversity of a given place.

The acknowledgement of their existence in language has been the focus of the Fungi Foundation, who started by formally defining the term,[31] and then teaming with conservation organisations to help them transition to 'mycologically inclusive language'. This 3F approach – fauna, flora, funga –

is leading the way for the formal recognition of fungi in conservation frameworks. It is up to all of us to say it: 'funga' means 'the diversity of fungi in a given place'.

This is all about recognising fungal protection as a vehicle for habitat protection. If we move towards implementing the 3F approach, the formal inclusion of fungi in conservation frameworks will help ensure an ecosystemic approach to conservation. This movement will positively impact the research, funding and political opportunities of fungi and mycology and advance the knowledge and protection of wider habitats.

Also, of course, we must 'rot on', letting the cycle of life reach decomposition by fungi, so all is recomposed.

Giuliana Furci is foundress and executive director of the Fungi Foundation. She is a Harvard University Associate, Dame of the Order of the Star of Italy, co-chair of the IUCN Fungal Conservation Committee, and the author of several titles, including a series of field guides to Chilean fungi and chapters in publications such as *Fantastic Fungi*. She is co-author of titles such as the first *State of the World's Fungi* (Kew, 2018) and *Biodiversidad de Chile: Patrimonio y Desafíos* (Ministry of the Environment, Chile, 2008). Giuliana has worked in the non-profit sector for the last seventeen years and has held consulting positions in US philanthropic foundations as well as full-time positions in international marine conservation NGOs, and Chilean environmental NGOs.

COPRINUS COMATUS

COMMON NAMES

Shaggy mane, shaggy ink cap, lawyer's wig

FAMILY	*Agaricaceae*
GENUS	*Coprinus*
SPECIES EPITHET	*comatus*

Coprinus comatus is a beginner-friendly mushroom to identify as it has no close look-alikes. Its long, cylindrical cap has a 'shaggy' texture due to the scales that crack and peel in warm weather. As it matures, the cap flares out at the edges and deliquesces – the fungus autodigests its own cap, which melts into a black, gooey liquid until only the white stipe is left. Mushrooms that deliquesce dissolve in a matter of hours. The spores are released in the liquid and carried away by wind. If you're feeling creative, you can collect the inky drops, mix them with essential oils and use them for writing or painting.

HISTORY AND CULTURE

C. comatus is a city mushroom. It thrives in lawns, city parks and golf courses, but its most famous location is when it breaks through concrete pavements. It absorbs water in its hyphae and explodes upwards with a force that is incredible for its size and fragility. It makes a heroic effort to continue its reproductive line.

PROPERTIES

EDIBLE

Yes, when young. Cook immediately after picking, before the cap and gills start to melt and turn black.

NUTRITIONAL PROFILE

A raw 100-gram serving contains 37 calories, composed of 90 % water, 8 % carbohydrate, 1 % protein and less than 1 % fat. Rich in vitamins and minerals, such as calcium and magnesium.

MEDICINAL

Yes, studies have shown anti-tumour, antioxidant, antimicrobial, antiviral, anti-inflammatory and anti-diabetic action.[32] Used traditionally in Asian countries to improve digestion and treat haemorrhoids.

PSYCHOACTIVE

No.

ENVIRONMENTAL REMEDIATION

Yes. Can accumulate toxic metals such as lead, arsenic and mercury in its sporing body. Useful as a bioindicator of soil pollution and in mycoremediation to sequester contaminants in the soil.[33]

SPORING BODY CHARACTERISTICS

CAP

2–6 cm wide
Cylindrical then bell-shaped
White and brown in centre
Covered with scales, becoming inky at edges

GILLS

White to pink then black
Crowded
Free from stipe

STIPE

6–30 cm tall
1–2 cm thick
Slightly widens at base
White
Ring at midpoint or base

SPORES

Black
Oval

FIELD DESCRIPTION

HABITAT

Seems to prefer artificial areas. Grows in clusters on disturbed soil, waste areas such as compost heaps, along roads and paths and even towards asphalt, often bursting through it.

DISTRIBUTION RANGE

Widespread in temperate and subtropical areas throughout Europe, North and South America, Australia, New Zealand, North Africa and Asia.

SEASON

Spring, summer and autumn.

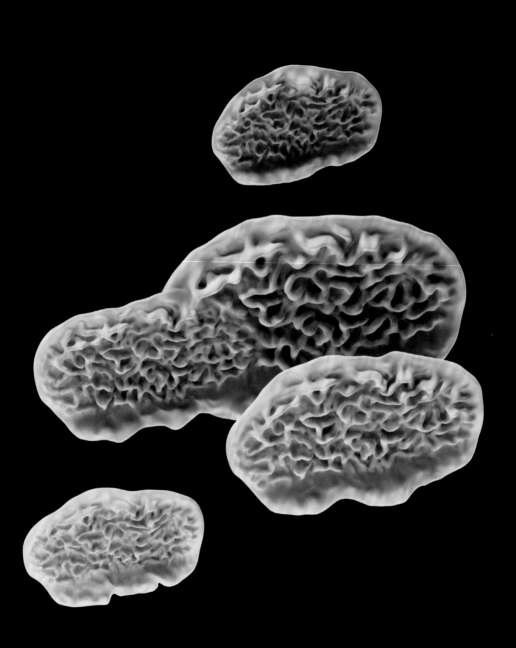

IRPEX LACTEUS

Irpex lacteus is saprophytic and spreads like a thick layer of white paint on the logs and branches it degrades. It has a porous surface and is a tooth fungus, similar to *Hericium erinaceus* (lion's mane), but instead of hanging spines, its teeth are blunt and short. At the edges, it can curl up and form a cap-like shelf. Its name is quite literal: *Irpex* means 'large rake with iron teeth' and *lacteus* means 'milky'.

HISTORY AND CULTURE

Due to its many forms, growth patterns and abundance, *I. lacteus* has been discovered, classified and rediscovered many times. Its classification is still extensively debated, but there is no doubt of its usefulness in mycoremediation. Its genome was sequenced in 2017 to understand the mechanisms behind its potent decomposition abilities. It is the model organism for researchers who are building a knowledge base of enzymes such as those produced by *I. lacteus*, which can be genetically engineered to degrade organic pollutants.[34]

PROPERTIES

EDIBLE

No.

MEDICINAL

Yes. Used traditionally in China to treat inflammation, bacterial and fungal infections and urinary retention.[35] Its polysaccharides are used in China as a clinical drug to cure chronic kidney inflammation.[36]

PSYCHOACTIVE

No.

ENVIRONMENTAL REMEDIATION

Yes, it is a white rot fungus. Well researched and known to degrade pollutants such as industrial effluents and pharmaceuticals.

SPORING BODY CHARACTERISTICS

BODY

Up to 3 cm thick
Irregularly shaped
May curl at edges to form cap
White-grey
Surface hairy, velvety

PORES

White to cream
2–3 per mm
If present, flat teeth up to 6 mm

SPORES

White
Oval

FIELD DESCRIPTION

HABITAT

Spreads across the decaying or dead wood of hardwood trees and occasionally coniferous trees.

DISTRIBUTION RANGE

North America.

SEASON

All year.

PLEUROTUS OSTREATUS

COMMON NAME

Oyster mushroom

FAMILY	*Pleurotaceae*
GENUS	*Pleurotus*
SPECIES EPITHET	*ostreatus*

Pleurotus ostreatus typically grows in thick bunches in striking colours, such as yellow, pink, brown, blue and grey. Pearl oysters are the most common – they are pure white with a pearl-like sheen. *Pleurotus* means 'sideways' in Latin, which appropriately describes this genus's unique horizontal growth angle, with its cap parallel to the ground like a shelf. *Ostreatus* means 'like the shell of an oyster' in Latin, which refers to the shape of the sporing body and perhaps its briny seafood flavour.

HISTORY AND CULTURE

Beginner cultivators can't go wrong with *P. ostreatus*. It has a voracious appetite and can be grown on coffee grounds, newspapers, wood chips and logs. It's even carnivorous: it evolved to excrete a chemical that knocks out nematodes that would otherwise feast on its sporing body. The mycelium envelops the worm like a lasso and consumes it. *P. ostreatus* was first cultivated in Germany due to food shortages in World War I and has since become a staple in diets across the world.

PROPERTIES

EDIBLE

Yes. Nutritious, delicious and versatile. Has a mild, nutty, seafood-like flavour with a meaty texture. Considered gourmet.

NUTRITIONAL PROFILE

A raw 100-gram serving contains 33 calories, composed of 89 % water, 6 % carbohydrate, 3 % protein and less than 1 % fat. Rich in vitamins, providing 20 % RDI of B vitamins, and minerals such as iron, zinc, potassium, phosphorus and selenium.

MEDICINAL

Yes. Naturally contains lovastatin, a cholesterol-lowering statin and the first statin approved by the FDA in 1987.

PSYCHOACTIVE

No.

ENVIRONMENTAL REMEDIATION

Yes, it is a white rot fungus. Extensively researched and tested in mycoremediation to break down oils, pesticides, herbicides and other industrial toxins. It can also accumulate heavy metals into its sporing body so that contaminants can be removed from soil or water.

SPORING BODY CHARACTERISTICS

CAP

2–20 cm wide
Oyster shell-shaped, convex, wavy
Ranges from white, grey and tan to dark brown
Surface smooth
Texture firm

GILLS

White or cream
Close or crowded
Run down stipe

STIPE

Usually absent or stubby
0.5–4 cm long and thick
White
Hairy at base

SPORES

White to lilac-grey
Oval

FIELD DESCRIPTION

HABITAT

Grows on decaying or dead hardwood trees, particularly beech, sycamore and aspen. Occasionally grows on conifers.

DISTRIBUTION RANGE

Widespread in temperate and subtropical forests. Common in Europe, Asia, Australia, New Zealand and North and South America.

SEASON

All year.

Fungi and the future

**MICHAEL LIM
YUN SHU**

It took fungi to wake us up to the natural world. True healing – of ourselves and the planet – can only occur by dissolving the walls that we erect and the labels we attach ourselves to. The belief that demarcates our physical bodies from the rest of the world is an illusion. Fungi show us this. They form intimate and indispensable relationships with all forms of life, permeating and supporting every ecosystem on Earth. Our planet as we know it would not exist without fungi. The kingdom of fungi radically challenges the concept of the individual.

As we journey through our one precious life, we are intimately intertwined with all life forms. Far from being separate from nature, we are an amalgamation of genes that have been mixing, mutating and transferring for billions of years. Each of us is an expression of life, to be explored and cherished.

We have been endowed with consciousness. We recognise this existence and our responsibility as stewards of our planet. There is so much that divides the world – changing it hinges on changing the way we perceive our place and purpose in it. To some, mould growing on milk is a nuisance; to others, it is an opportunity to produce cheese. Fungi can offer endless solutions to our physical, mental and spiritual crises. If we can change our mind about a fungus, imagine what else we can change.

Mycology needs more scientists. More funding. More activism. More protection. More people celebrating fungi through different mediums. The next time you eat mushrooms, drink wine or take a walk through the forest, give a moment of gratitude to this humble kingdom.

This is only the beginning of our existential awakening. Fungi is just one of many thrilling explorations. Science, philosophy, art, mathematics and many more disciplines offer new dimensions of mental models for us to understand and appreciate the macrocosm. We have eyes to take in sights, noses to take in scents and bodies to deeply immerse into nature. Go out and experience the wonder for yourself.

This is a chance to raise our consciousness and it's in the pursuit of these greater goals that the human spirit shines through. When a flash of iridescent knowing arises, follow it. These inquiries change you. They should change you. We are capable of so much more connection, empathy and creativity.

May we loosen our grip on dogmas, question our old assumptions and embrace new ideas. May we consume consciously, give generously, choose love over fear and have fun along the way. We are travellers on a cosmic trip.

It is a pleasure to have encountered, loved and shared this time in space with you.

Endnotes

FOREWORD

1. Nina Graboi, *One Foot in the Future: A Woman's Spiritual Journey*, Aerial Press, 1991, p. 164.

KINGDOM OF FUNGI

1. Anne Casselman, 'Strange but true: The largest organism on Earth is a fungus', *Scientific American*, 4 October 2007, <scientificamerican.com/article/strange-but-true-largest-organism-is-fungus>.

2. Patrick Forterre, 'The universal tree of life: An update', *Frontiers in Microbiology*, vol. 6, 2015, <doi.org/10.3389/fmicb.2015.00717>.

3. Petr Baldrian, Tomáš Větrovský, Clémentine Lepinay and Petr Kohout, 'High-throughput sequencing view on the magnitude of global fungal diversity', *Fungal Diversity*, 2021, <doi.org/10.1007/s13225-021-00472-y>.

4. Unai Ugalde and Ana Belén Rodriguez-Urra, '9 Autoregulatory signals in mycelial fungi', *The Mycota*, vol. 1, 2016, <doi.org/10.1007/978-3-319-25844-7_9>.

5. Daniel S Heckman, David M Geiser, Brooke R Eidell, Rebecca L Stauffer, Natalie L Kardos and S Blair Hedges, 'Molecular evidence for the early colonization of land by fungi and plants', *Science*, vol. 293, issue 5532, 2001, <doi.org/10.1126/science.1061457>.

6. Claire P Humphreys, Peter J Franks, Mark Rees, Martin I Bidartondo, Jonathan R Leake and David J Beerling, 'Mutualistic mycorrhiza-like symbiosis in the most ancient group of land plants', *Nature Communications*, no. 1, article 103, 2010, <doi.org/10.1038/ncomms1105>.

7. François Lutzoni, Michael D Nowak, Michael E Alfaro, Valérie Reeb, Jolanta Miadlikowska, Michael Krug, A Elizabeth Arnold, Louise A Lewis, David L Swofford, David Hibbett, Khidir Hilu, Timothy Y James, Dietmar Quandt and Susana Magallón, 'Contemporaneous radiations of fungi and plants linked to symbiosis', *Nature Communications*, no. 9, article 5451, 2018, <doi.org/10.1038/s41467-018-07849-9>.

8. Michael Krings, Carla J Harper and Edith L Taylor, 'Fungi and fungal interactions in the Rhynie chert: A review of the evidence, with the description of *Perexiflasca tayloriana*', *Philosophical Transactions of The Royal Society B*, vol. 373, issue 1739, 2018, <doi.org/10.1098/rstb.2016.0500>.

9. Eun-Hwa Lee, Ju-Kyeong Eo, Kang-Hyeon Ka and Ahn-Heum Eom, 'Diversity of arbuscular mycorrhizal fungi and their roles in ecosystems', *Mycobiology*, vol. 41, issue 3, 2013, <doi.org/10.5941/MYCO.2013.41.3.121>.

10. Mark C Brundrett and Leho Tedersoo, 'Evolutionary history of mycorrhizal symbioses and global host plant diversity', *New Phytologist*, vol. 220, issue 4, 2018, <doi.org/10.1111/nph.14976>.

11. Johan Asplund and David A Wardle, 'How lichens impact on terrestrial community and ecosystem properties', *Biological Reviews*, vol. 92, issue 3, 2016, <doi.org/10.1111/brv.12305>.

12. Matthew Phelan, 'Why fungi adapt so well to life in space', *Scienceline*, 7 March 2018, <scienceline.org/2018/03/fungi-love-to-grow-in-outer-space>.

13. Anderson G Oliveira, Cassius V Stevani, Hans E Waldenmaier, Vadim Viviani Jillian M Emerson, Jennifer J Loros and Jay C Dunlap, 'Circadian control sheds light on fungal bioluminescence', *Current Biology*, vol. 25, issue 7, 2015, <doi.org/10.1016/j.cub.2015.02.021>.

14. Suzanne W Simard, David A Perry, Melanie D Jones, David D Myrold, Daniel M Durall and Randy Molina, 'Net transfer of carbon between ectomycorrhizal tree species in the field', *Nature*, vol. 388, 1997, <doi.org/10.1038/41557>.

15. Ferris Jabr, 'A vast, ancient and intricate society: The secret social network of old-growth forests', *Sydney Morning Herald*, 29 January 2021, <smh.com.au/environment/sustainability/a-vast-ancient-and-intricate-society-the-secret-social-network-of-old-growth-forests-20200703-p558ti.html>.

16. Anouk van't Padje, Loreto Oyarte Galvez, Malin Klein, Mark A Hink, Marten Postma, Thomas Shimizu and E Toby Kiers, 'Temporal tracking of quantum-dot apatite across in vitro mycorrhizal networks shows how host demand can influence fungal nutrient transfer strategies', *The ISME Journal*, vol. 15, 2021, <doi.org/10.1038/s41396-020-00786-w>; Anouk van't Padje, Gijsbert DA Werner and E Toby Kiers, 'Mycorrhizal fungi control phosphorus value in trade symbiosis with host roots when exposed to abrupt "crashes" and "booms" of resource availability', *New Phytologist*, vol. 229, issue 5, 2020, <doi.org/10.1111/nph.17055>.

17. Toby Kiers, 'Lessons from fungi on markets and economics' [video], *TED*, September 2019, <ted.com/talks/toby_kiers_lessons_from_fungi_on_markets_and_economics>.

FOOD

1. Russell F Doolittle, Da-Fei Feng, Simon Tsang, Glen Cho and Elizabeth Little, 'Determining divergence times of the major kingdoms of living organisms with a protein clock', *Science*, vol. 271, issue 5248, 1996, <doi.org/10.1126/science.271.5248.470>.

2. University of Pennsylvania, '9,000-year history of Chinese fermented beverages confirmed', *ScienceDaily*, 7 December 2004, <sciencedaily.com/releases/2004/12/041206205817.htm>.

3. Anahita Shams, 'Does Shiraz wine come from Iran?', *BBC*, 3 February 2017, <bbc.com/news/world-middle-east-38771806>.

4. Vera Meyer, Evelina Y Basenko and Han AB Wösten, 'Growing a circular economy with fungal biotechnology: A white paper', *Fungal Biology and Biotechnology*, no. 7, 2020, <doi.org/10.1186/s40694-020-00095-z>.

5. Pau Loke Show, Kehinde Opeyemi Oladele, Qi Yan Siew, Fitri Abdul Aziz Zakry, John Chi-Wei Lan and Tau Chuan Ling, 'Overview of citric acid production from *Aspergillus niger*', *Frontiers in Life Science*, vol. 8, no. 3, 2015, <doi.org/10.1080/21553769.2015.1033653>.

6. William Shurtleff and Akiko Aoyagi, *History of Tempeh and Tempeh Products (1815–2020): Bibliography and Sourcebook*, Soyinfo Center, Lafayette, 2020, p. 351.

7. Marianna Cerini, 'Tempeh, Indonesia's wonder food', *The Economist*, 23 January 2020, <economist.com/1843/2020/01/23/tempeh-indonesias-wonder-food>.

8. @david_zilber, 'Biomimicry is a fascinating way...' [Instagram post], David Chaim Jacob Zilber, 26 May 2020, <instagram.com/p/CAptR8qpN-T>.

9. Winston Churchill and Steven Spurrier, 'Fifty years hence', *Strand Magazine*, issue 82, no. 49, 1931.

10. Eben Bayer, interview with the authors, 2020.

11. Mary Jo Feeney, Amy Myrdal Miller and Peter Roupas, 'Mushrooms—biologically distinct and nutritionally unique', *Nutrition Today*, vol. 49, issue 6, 2014, <journals.lww.com/nutritiontodayonline/toc/2014/11000>.

12. National Institutes of Health, *Selenium – Fact Sheet for Health Professionals*, US Department of Health & Human Services, 26 March 2021, <ods.od.nih.gov/factsheets/Selenium-HealthProfessional>.

13. National Institutes of Health, *Potassium – Fact Sheet for Health Professionals*, US Department of Health & Human Services, 26 March 2021, <ods.od.nih.gov/factsheets/Potassium-HealthProfessional>.

14. National Institutes of Health, *Phosphorus – Fact Sheet for Health Professionals*, US Department of Health & Human Services, 26 March 2021, <ods.od.nih.gov/factsheets/Phosphorus-HealthProfessional>.

15. National Institutes of Health, *Folate – Fact Sheet for Health Professionals*, US Department of Health & Human Services, 29 March 2021, <ods.od.nih.gov/factsheets/Folate-HealthProfessional>.

16. National Institutes of Health, *Zinc – Fact Sheet for Health Professionals*, US Department of Health & Human Services, 26 March 2021, <ods.od.nih.gov/factsheets/Zinc-HealthProfessional>.

17. Environmental Protection Agency, *Global Greenhouse Gas Emissions Data*, United States Environmental Protection Agency, n.d., <epa.gov/ghgemissions/global-greenhouse-gas-emissions-data>.

18. IU = International Unit, which is a measure of biological activity that is different for each substance.

19. Mary Jo Feeney et al., 'Mushrooms—biologically distinct and nutritionally unique'.

20. Victor L Fulgoni III and Sanjiv Agarwal, 'Nutritional impact of adding a serving of mushrooms on usual intakes and nutrient adequacy using National Health and Nutrition Examination Survey 2011–2016 data', *Food Science and Nutrition*, vol. 9, issue 3, 2021, <doi.org/10.1002/fsn3.2120>.

21. Sonya Sachdeva, Marla R Emery and Patrick T Hurley, 'Depiction of wild food foraging practices in the media: Impact of the great recession', *Society & Natural Resources*, vol. 31, issue 8, 2018, <doi.org/10.1080/08941920.2018.1 450914>.

22. Simon Egli, Martina Peter, Christoph Buser, Werner Stahel and François Ayer, 'Mushroom picking does not impair future harvests – results of a long-term study in Switzerland', *Biological Conservation*, vol. 129, issue 2, 2006, <doi.org/10.1016/j.biocon.2005.10.042>.

23. J Avinash, S Vinay, Kunal Jha, Diptajit Das, BS Goutham and Gunjan Kumar, 'The unexplored anticaries potential of shiitake mushroom', *Pharmacognosy Reviews*, vol. 10, issue 20, 2016, <doi.org/10.4103/0973-7847.194039>.

24. Shwet Kamal, VP Sharma, Mamta Gupta, Anupam Barh and Manjit Singh, 'Genetics and breeding of white button mushroom, *Agaricus bisporus* (Lange.) Imbach.– A comprehensive review', *Mushroom Research*, vol. 28, no. 1, 2019, <doi.org/10.36036/MR.28.1.2019.91938>.

25. Bozena Muszyńska, Katarzyna Kała, Anna Firlej and Katarzyna Sułkowska-Ziaja, '*Cantharellus cibarius* – Culinary –Medicinal mushroom content and biological activity', *Acta poloniae pharmaceutica*, vol. 73, issue 3, 2016, <pubmed.ncbi.nlm.nih.gov/27476275>.

26. Sibel Yildiz, Aysenur Gurgen and Ugur Cevik, 'Accumulation of metals in some wild and cultivated mushroom species', *Sigma Journal of Engineering and Natural Science*, vol. 37, issue 4, 2019, <researchgate.net/publication/338254406_accumulation_of_metals_in_some_wild_and_cultivated_mushroom_species>.

27. Michel Feussi Tala, Jianchun Qin, Joseph T Ndongo and Hartmut Laatsch, 'New azulene-type sesquiterpenoids from the fruiting bodies of *Lactarius deliciosus*', *Natural Products and Bioprospecting*, vol. 7, issue 3, 2017, <doi.org/10.1007/s13659-017-0130-1>.

28. Hasan Akgül, Mustafa Sevindik, Celal Bal, Hayri Baba and Zeliha Selamoglu, 'Medical properties of edible mushroom *Lactarius deliciosus*', *Biological Activities of Mushrooms*, 2019, <researchgate.net/publication/336773973_medical_properties_of_edible_mushroom_lactarius_deliciosus>.

29. Alexis Guerin-Laguette, Claude Plassard and Daniel Mousain, 'Effects of experimental conditions on mycorrhizal relationships between *Pinus sylvestris* and *Lactarius deliciosus* and unprecedented fruit-body formation of the saffron milk cap under controlled soilless conditions', *Canadian Journal of Microbiology*, vol. 46, no. 9, 2000, <doi.org/10.1139/w00-059>.

30. Weldesemayat Gorems Woldemariam, 'Mushrooms in the bio-remediation of wastes from soil', *Advances in Life Science and Technology*, vol. 76, 2019, <doi.org/10.7176/ALST/76-04>.

31. Jeong-Ah Kim, Edward Lau, David Tay and Esperanza J Carcache De Blanco, 'Antioxidant and NF-κB inhibitory constituents isolated from *Morchella esculenta*', *Natural product research*, vol. 25, issue 15, 2011, <doi.org/10.1080/14786410802425746>; B Nitha and KK Janardhanan, 'Aqueous-ethanolic extract of morel mushroom mycelium *Morchella esculenta*, protects cisplatin and gentamicin induced nephrotoxicity in mice', *Food and Chemical Toxicology*, vol. 46, issue 9, 2008, <doi.org/10.1016/j.fct.2008.07.007>.

32. Yiling Hou, Xiang Ding, Wanru Hou, Jie Zhong, Hongqing Zhu, Binxiang Ma, Ting Xu and Junhua Li, 'Anti-microorganism, anti-tumor, and immune activities of a novel polysaccharide isolated from *Tricholoma matsutake*', *Pharmacognosy Magazine*, vol. 9, issue 35, 2013, <doi.org/10.4103/0973-1296.113278>.

MEDICINE

1. Walter Kutschera and Werner Rom, 'Ötzi, the prehistoric iceman', *Nuclear Instruments and Methods in Physics Research Section B: Beam Interactions with Materials and Atoms*, vol. 164–165, 2000, <doi.org/10.1016/S0168-583X(99)01196-9>.

2. Sissi Wachtel-Galor, John Yuen, John A Buswell, and Iris FF Benzie, '*Ganoderma lucidum* (lingzhi or reishi): A medicinal mushroom', *Herbal Medicine: Biomolecular and Clinical Aspects*, 2nd edition, 2011, chapter 9, <pubmed.ncbi.nlm.nih.gov/22593926/>.

3. Royal Botanic Gardens Kew, *State of the World's Fungi 2018* [website], 2018, <stateoftheworldsfungi.org>.

4. Alexander N Shikov, Olga N Pozharitskaya, Valer G Makarov, Hildebert Wagner, Rob Verpoorte and Michael Heinrich, 'Medicinal plants of the Russian pharmacopoeia; their history and applications', *Journal of Ethnopharmacology*, vol. 154, issue 3, 2014, <doi.org/10.1016/j.jep.2014.04.007>.

5. Howard Markel, 'The real story behind penicillin', *News Hour*, PBS, 27 September 2013, <pbs.org/newshour/health/the-real-story-behind-the-worlds-first-antibiotic>.

6. 'The Nobel Prize in Physiology or Medicine 1945', The Nobel Prize, 2021, <nobelprize.org/prizes/medicine/1945/summary>.

7. Siang Yong Tan and Yvonne Tatsumura, 'Alexander Fleming (1881–1955): Dicoverer of penicillin', Singapore Medical Journal, vol. 56, issue 7, July 2015, <doi.org/10.11622/smedj.2015105>.

8. Sean P Gordon, Elizabeth Tseng, Asaf Salamov, Jiwei Zhang, Xiandong Meng, Zhiying Zhao, Dongwan Kang, Jason Underwood, Igor V Grigoriev, Melania Figueroa, Jonathan S Schilling, Feng Chen and Zhong Wang, 'Widespread polycistronic transcripts in fungi revealed by single-molecule mRNA sequencing', *PLoS ONE*, vol. 7, issue 10, 2015, <doi.org/10.1371/journal.pone.0132628>.

9. Paul Stamets and Heather Zwickey, 'Medicinal mushrooms: Ancient remedies meet modern science', *Integrative Medicine (Encinitas)*, vol. 13, 2014, <pubmed.ncbi.nlm.nih.gov/26770081>.

10. Jeff Chilton, interview with the authors, 2020.

11. Koichiro Mori, Satoshi Inatomi, Kenzi Ouchi, Yoshihito Azumi and Takashi Tuchida, 'Improving effects of the mushroom yamabushitake (*Hericium erinaceus*) on mild cognitive impairment: A double-blind placebo-controlled clinical trial', *Phytotherapy Research*, vol. 23, issue 3, 2008, <doi.org/10.1002/ptr.2634>.

12. Yuusuke Saitsu, Akemi Nishide, Kenji Kikushima, Kuniyoshi Shimizu and Koichiro Ohnuki, 'Improvement of cognitive functions by oral intake of *Hericium erinaceus*', *Biomedical Research*, vol. 40, issue 4, 2019, <doi.org/10.2220/biomedres.40.125>.

13. Koichiro Mori, Yutaro Obara, Mitsuru Hirota, Yoshihito Azumi, Satomi Kinugasa, Satoshi Inatomi and Norimichi Nakahata, 'Nerve growth factor-inducing activity of *Hericium erinaceus* in 1321N1 human astrocytoma cells', *Biological and Pharmaceutical Bulletin*, vol. 31, issue 9, 2008, <doi.org/10.1248/bpb.31.1727>.

14. Keenan A Walker, Rebecca F Gottesman, Aozhou Wu, David S Knopman, Alden L Gross, Thomas H Mosley Jr, Elizabeth Selvin and B Gwen Windham, 'Systemic inflammation during midlife and cognitive change over 20 years: The ARIC study', *Neurology*, vol. 92, 2019, <pubmed.ncbi.nlm.nih.gov/30760633>.

15. Jeff Chilton, interview with the authors, 2020.

16. Katie R Hirsch, Abbie E Smith-Ryan, Erica J Roelofs, Eric T Trexler and Meredith G Mock, '*Cordyceps militaris* improves tolerance to high intensity exercise after acute and chronic supplementation', *Journal of Dietary Supplements*, vol. 14, issue 1, 2016, <doi.org/10.1080/19390211.2016.1203386>.

17. Steve Chen, Zhaoping Li, Robert Krochmal, Marlon Abrazado, Woosong Kim and Christopher B Cooper, 'Effect of Cs-4® (*Cordyceps sinensis*) on exercise performance in healthy older subjects: A double-blind, placebo-controlled trial', *The Journal of Alternative and Complementary Medicine*, vol. 16, no. 5, 2010, <doi.org/10.1089/acm.2009.0226>.

18. Kanitta Jiraungkoorskul and Wannee Jiraungkoorskul, 'Review of naturopathy of medical mushroom, *Ophiocordyceps sinensis*, in sexual dysfunction', *Pharmacognosy Reviews*, vol. 10, issue 19, 2016, <doi.org/10.4103/0973-7847.176566>.

19. Parris M Kidd, 'The use of mushroom glucans and proteoglycans in cancer treatment', *Alternative Medicine Review: a Journal of Clinical Therapeutic*, vol. 5, issue 1, 2000, <pubmed.ncbi.nlm.nih.gov/10696116>.

20. Parris M Kidd, 'The use of mushroom glucans and proteoglycans in cancer treatment'.

21. Stanford Medicine, 'Yeast engineered to manufacture complex medicine', *ScienceDaily*, 2 April 2018, <sciencedaily.com/releases/2018/04/180402192627.htm>.

22. Barry V McCleary and Anna Draga, 'Measurement of ß-glucan in mushrooms and mycelial products', *Journal of AOAC International*, vol. 99, issue 2, 2016, <doi.org/10.5740/jaoacint.15-0289>.

23. Jeff Chilton, interview with the authors, 2020.

24. Ayodele Rotimi Ipeaiyeda, Clementina Oyinkansola Adenipekun and Oluwatola Oluwole, 'Bioremediation potential of *Ganoderma lucidum* (Curt:Fr) P. Karsten to remove toxic metals from abandoned battery slag dumpsite soil and immobilisation of metal absorbed fungi in bricks', *Cogent Environmental Science*, vol. 6, issue 1, 2020, <doi.org/10.1080/23 311843.2020.1847400>.

25. Nahid Akhtara and M Amin-ul Mannan, 'Mycoremediation: Expunging environmental pollutants', *Biotechnology Reports*, vol. 26, 2020, <doi.org/10.1016/j.btre.2020.e00452>.

PSYCHEDELICS

1. RR Griffiths, WA Richards, U McCann and R Jesse, 'Psilocybin can occasion mystical-type experiences having substantial and sustained personal meaning and spiritual significance', *Psychopharmacology*, vol. 187, 2006, <doi.org/10.1007/s00213-006-0457-5>.

2. Kerry Lotzof, *Are we really made of stardust?*, Natural History Museum website, n.d., <nhm.ac.uk/discover/are-we-really-made-of-stardust.html>.

3. Elizabeth Howell, *Humans really are made of stardust, and a new study proves it*, Space.com, 2017, <space.com/35276-humans-made-of-stardust-galaxy-life-elements.html>.

4. Cynthia Carson Bisbee, *Psychedelic Prophets: The Letters of Aldous Huxley and Humphry Osmond*, McGill-Queen's University Press, Montreal, 2018, p. 267.

5. Carl Sagan and Ann Druyan, *The Demon-Haunted World: Science as a Candle in the Dark*, Random House, New York, 1995.

6. Jamilah R George, Timothy I Michaels, Jae Sevelius and Monnica T Williams, 'The psychedelic renaissance and the limitations of a White-dominant medical framework: A call for indigenous and ethnic minority inclusion', *Journal of Psychedelic Studies*, vol. 4, issue 1, 2020, <doi.org/10.1556/2054.2019.015>.

7. Bianca M Dinkelaar, 'Plato and the language of mysteries', *Mnemosyne*, vol. 73, issue 1, 2020, <doi.org/10.1163/1568525X-12342654>

8. David Horan, 'Plato's *Phaedrus*', *The Dialogues of Plato*, 2008, <platonicfoundation.org/phaedrus>.

9. Alexander Shulgin and Ann Shulgin, *TiHKAL The Continuation*, Transform Press, Berkeley, 1997.

10. Antonia Tripolitis, *Religions of the Hellenistic-Roman Age*, Eerdmans Publishing Company, 2001.

11. FJ Carod-Artal, 'Hallucinogenic drugs in pre-Columbian Mesoamerican cultures', *Neurología*, vol. 30, issue 1, 2015, <doi.org/10.1016/j.nrl.2011.07.003>.

12. Richard Evans Schultes, Albert Hofmann and Christian Rätsch, *Plants of the Gods, Inner Traditions*, Rochester, 1979.

13. Richard Evans Schultes, 'The identification of teonanácatl, a narcotic basidiomycete of the Aztecs', *Botanical Museum Leaflets of Harvard University – Plantae Mexicanae II*, vol. 7, no. 3, 1939, <samorini.it/doc1/alt_aut/sz/schultes-identification-of-teonanacatl.pdf>.

14. Albert Hofmann, *LSD, My Problem Child*, Oxford University Press, Oxford, 2013, p. 18.

15. Albert Hofmann, *LSD, My Problem Child*, p. 19.

16. David E Nichols, 'Psychedelics', *Pharmacological Reviews*, vol. 68, issue 2, 2016, <doi.org/10.1124/pr.115.011478>.

17. R Gordon Wasson, 'Seeking the Magic Mushroom', *LIFE*, 1957.

18. R Gordon Wasson, 'Seeking the Magic Mushroom', *LIFE*.

19. Lyrics from the song 'Maria Sabina' by El Tri, Producciones Lora, Mexico City, 1989.

20. Select Committee on Intelligence and Committee on Human Resources, *Project MK-ULTRA, The CIA's Program of Research in Behavioral Modification*, 1977, <intelligence.senate.gov/sites/default/files/hearings/95mkultra.pdf>.

21. Ram Dass, *Be Here Now*, Lama Foundation, San Cristobal, 1971.

22. Ram Dass, *Be Here Now*.

23. Walter Norman Pahnke, *Drugs and Mysticism*, Harvard University, 1963, <maps.org/images/pdf/books/pahnke/walter_pahnke_drugs_and_mysticism.pdf>.

24. San Luis Obispo, 'Timothy Leary, drug advocate, walks away from coast prison', *The New York Times*, 14 September 1970, <nytimes.com/1970/09/14/archives/timothy-leary-drug-advocate-walks-away-from-coast-prison.html>.

25. Rob Harper (director), *Journeys to the Edge of Consciousness*, Hidden Depths Production, 2019, <journeysmovie.com>.

26. Terence McKenna, 'McNature', *Psychedelic Salon* [podcast], 16 September 2009, <psychedelicsalon.com/podcast-197-mcnature>.

27. Walter Isaacson, *Steve Jobs*, Little, Brown, London, 2011.

28. RR Griffiths, WA Richards, U McCann and R Jesse, 'Psilocybin can occasion mystical-type experiences having substantial and sustained personal meaning and spiritual significance', *Psychopharmacology*, vol. 187, 2006, <doi.org/10.1007/s00213-006-0457-5>.

29. Kevin Balktick (director), *The Johns Hopkins Story*, Horizons Media, <horizons.nyc/films/john-hopkins-story>.

30. FX Vollenweider, MF Vollenweider-Scherpenhuyzen, A Bäbler, H Vogel and D Hell, 'Psilocybin induces schizophrenia-like psychosis in humans via a serotonin-2 agonist action', *NeuroReport*, vol. 9, issue 17, 1998, <doi.org/10.1097/00001756-199812010-00024>.

31. Robin L Carhart-Harris, David Erritzoe, Tim Williams, James M Stone, Laurence J Reed, Alessandro Colasanti, Robin J Tyacke, Robert Leech, Andrea L Malizia, Kevin Murphy, Peter Hobden, John Evans, Amanda Feilding, Richard G Wise, and David J Nutt, 'Neural correlates of the psychedelic state as determined by fMRI studies with psilocybin', *PNAS*, vol. 109, no. 6, 2012, <doi.org/10.1073/pnas.1119598109>.

32. Cola SL Lo, Samuel MY Ho and Steven D Hollon, 'The effects of rumination and negative cognitive styles on depression: A mediation analysis', *Behaviour Research and Therapy*, vol. 46, issue 4, 2008, <doi.org/10.1016/j.brat.2008.01.013>.

33. G Petri, P Expert, F Turkheimer, R Carhart-Harris, D Nutt, PJ Hellyer and F Vaccarino, 'Homological scaffolds of brain functional networks', *Journal of the Royal Society Interface*, 2014, <doi.org/10.1098/rsif.2014.0873>.

34. Stanislav Grof, *LSD Psychotherapy: The Healing Potential of Psychedelic Medicine*, Multidisciplinary Association for Psychedelic Studies, San Jose, 2008.

35. Image adapted from G Petri, P Expert, F Turkheimer, R Carhart-Harris, D Nutt, PJ Hellyer and F Vaccarino, 'Homological scaffolds of brain functional networks', *Journal of the Royal Society Interface*, 2014, <doi.org/10.1098/rsif.2014.0873>.

36. David Jay Brown, *Frontiers of Psychedelic Consciousness*, Park Street Press, Rochester, 2015.

37. Prof David J Nutt, Leslie A King and Lawrence D Phillips, 'Drug harms in the UK: A multicriteria decision analysis', *The Lancet*, vol. 376, no. 9572, 2010, <doi.org/10.1016/S0140-6736(10)61462-6>.

38. Matthew W Johnson, William A Richards and Roland R Griffiths, 'Human hallucinogen research: Guidelines for safety', *Journal of Psychopharmacology*, vol. 22, issue 6, 2008, <doi.org/10.1177/0269881108093587>.

39. Alan Watts, *The Joyous Cosmology: Adventures in the Chemistry of Consciousness*, Vintage Books, New York, 1962.

40. Catherine K Ettman, Salma M Abdalla, Gregory H Cohen, Laura Sampson, Patrick M Vivier and Sandro Galea, 'Prevalence of depression symptoms in US adults before and during the COVID-19 pandemic', *JAMA Network Open*, vol. 3, issue 9, 2020, <doi.org/10.1001%2Fjamanetworkopen.2020.19686>.

41. Robin Carhart-Harris, Bruna Giribaldi, Rosalind Watts and Michelle Baker-Jones, 'Trial of psilocybin versus escitalopram for depression', *The New England Journal of Medicine*, 2021, <doi.org/10.1056/NEJMoa2032994>.

42. Marshall Tyler, interview with the authors.

43. Christopher G Hudson, 'Socioeconomic status and mental illness: Tests of the social Causation and selection hypotheses', *American Journal of Orthopsychiatry*, vol. 75, no. 1, 2005, <doi.org/10.1037/0002-9432.75.1.3>.

44. *North Star Ethics Pledge*, North Star, 2020, <northstar.guide/ethicspledge>.

45. Rick Doblin, correspondence with the authors, 2020.

46. *Denver, Colorado, Initiated Ordinance 301, Psilocybin Mushroom Initiative (May 2019)*, Ballotpedia, n.d., <ballotpedia.org/Denver,_Colorado,_Initiated_Ordinance_301,_Psilocybin_Mushroom_Initiative_(May_2019)>.

47. Tim Ferriss, 'An urgent plea to users of psychedelics: Let's consider a more ethical menu of plants and compounds', *The Tim Ferriss Show*, 21 February 2021, <tim.blog/2021/02/21/urgent-plea-users-of-psychedelics-ethical-plants-compounds>.

48. Abraham H Maslow, *Motivation and Personality*, Harpers, New York, 1954, p. 93.

49. Abraham H Maslow, *Religions, Values, and Peak-Experiences*, Ohio State University Press, Columbus, 1944, p. 27.

50. Alan Watts, *The Book on the Taboo Against Knowing Who You Are*, Pantheon Books, New York, 1966.

51. Małgorzata Drewnowska, Krzysztof Lipka, Grażyna Jarzyńska, Dorota Danisiewicz-Czupryńska and Jerzy Falandysz, 'Investigation on metallic elements in fungus *Amanita muscaria* (fly agaric) and the forest soils from the Mazurian lakes district of Poland', *Fresenius Environmental Bulletin*, vol. 22, no. 2, 2013, <researchgate.net/profile/Jerzy-Falandysz/publication/285839446_Investigation_on_metallic_elements_in_fungus_Amanita_muscaria_Fly_Agaric_and_the_forest_soils_from_the_Mazurian_Lakes_District_of_Poland/links/5a032bdaaca2720c32676ff0/Investigation-on-metallic-elements-in-fungus-Amanita-muscaria-Fly-Agaric-and-the-forest-soils-from-the-Mazurian-Lakes-District-of-Poland.pdf>.

ENVIRONMENT

1. William J Ripple, Christopher Wolf, Thomas M Newsome, Phoebe Barnard and William R Moomaw, 'World scientists' warning of a climate emergency', *BioScience*, vol. 70, issue 1, 2020, <doi.org/10.1093/biosci/biz152>.

2. Peter H Gleick, *Water in Crisis: A Guide to the World's Fresh Water Resources*, Oxford University Press, Oxford, 1993.

3. UNESCO World Water Assessment Programme, *The United Nations World Water Development Report 2017: Wastewater: The UntappedRresource; Facts and Figures*, UNESCO, 2017, <unesdoc.unesco.org/ark:/48223/pf0000247553>.

4. Tradd Cotter, interview with the authors, 2020.

5. KE Clemmensen, A Bahr, O Ovaskainen, A Dahlberg, A Ekblad, H Wallander, J Stenlid, RD Finlay, DA Wardle, and BD Lindahl, 'Roots and associated fungi drive long-term carbon sequestration in boreal forest', *Science*, vol. 339, issue 6127, 2013, <doi.org/10.1126/science.1231923>.

6. Kathleen K Treseder and Sandra R Holden, 'Fungal carbon Sequestration', *Science*, vol. 339, issue 6127, 2013, <doi.org/10.1126/science.1236338>.

7. Jeff Ravage, interview with the authors, 2020.

8. Levon Durr, *Mycoremediation Project: Using Mycelium to Clean Up Diesel-Contaminated Soil in Orleans, California*, Fungaia Farm, 2016, <fungaiafarm.com/wp-content/uploads/2014/07/MycoremediationReport_FungaiaFarm_2016.pdf>.

9. Levon Durr, interview with the authors, 2020.

10. Joanne Rodriguez, interview with the authors, 2020.

11. Roland Geyer, Jenna R Jambeck and Kara Lavender Law, 'Production, use, and fate of all plastics ever made', *Science Advances*, vol. 3, no. 7, 2017, <doi.org/10.1126/sciadv.1700782>.

12. Marcus Eriksen, Laurent CM Lebreton, Henry S Carson, Martin Thiel, Charles J Moore, Jose C Borerro, Francois Galgani, Peter G Ryan and Julia Reisser, 'Plastic pollution in the world's oceans: More than 5 trillion plastic pieces weighing over 250,000 tons afloat at sea', *PLOS*, 2014, <doi.org/10.1371/journal.pone.0111913>.

13. Sehroon Khan, Sadia Nadir, Zia Ullah Shah, Aamer Ali Shah, Samantha C Karunarathna, Jianchu Xu, Afsar Khan, Shahzad Munir and Fariha Hasan, 'Biodegradation of polyester polyurethane by *Aspergillus tubingensis*', *Environmental Pollution*, vol. 225, 2017, <doi.org/10.1016/j.envpol.2017.03.012>.

14. Food and Agriculture Organization of the United Nations, 'New standards to curb the global spread of plant pests and diseases', Food and Agriculture Organization of the United Nations, 2019, <fao.org/news/story/en/item/1187738/icode>.

15. The Business Research Company, *Pesticide and Other Agricultural Chemicals Global Market Report 2021: COVID-19 Impact and Recovery to 2030*, The Business Research Company, 2021, <researchandmarkets.com/reports/5240332/pesticide-and-other-agricultural-chemicals-global>.

16. Kristin S Schafer and Emily C Marquez, *A Generation in Jeopardy: How Pesticides Are Undermining Our Children's Health & Intelligence*, Pesticide Action Network North America, 2013.

17. Environmental Protection Agency, *What are biopesticides?*, United States Environmental Protection Agency, n.d., <epa.gov/ingredients-used-pesticide-products/what-are-biopesticides>.

18. 'How Dangerous Is Pesticide Drift?', *Scientific American*, 2012, <scientificamerican.com/article/pesticide-drift>.

19. Stephen Leahy, 'World Water Day: The cost of cotton in water-challenged India', *The Guardian*, 21 March 2015, <theguardian.com/sustainable-business/2015/mar/20/cost-cotton-water-challenged-india-world-water-day>.

20. Freek Appels and Han Wosten, 'Mycelium materials', *Encyclopedia of Mycology*, vol. 2, 2021, <doi.org/10.1016/B978-0-12-809633-8.21131-X>.

21. World Wild Life, *Deforestation and Forest Degradation*, World Wild Life, n.d., <worldwildlife.org/threats/deforestation-and-forest-degradation>.

22. Maurizio Montalti, interview with the authors, 2020.

23. Maurizio Montalti, interview with the authors, 2020.

24. Ehab Sayed, interview with the authors, 2020.

25. HAB Woesten, P Krijgsheld, M Montalti, H Lakk, 'Growing fungi structures in space', *European Space Agency, the Advanced Concepts Team, Ariadna Final Report*, European Space Agency, 2018, <esa.int/gsp/ACT/doc/ARI/ARI%20Study%20Report/ACT-RPT-HAB-ARI-16-6101-Fungi_structures.pdf>.

26. Lisa Anne Hamilton and Steven Feit, *Plastic & Climate: The Hidden Costs of a Plastic Planet*, Center for International Environmental Law, 2019, <ciel.org/reports/plastic-health-the-hidden-costs-of-a-plastic-planet-may-2019>.

27. Eben Bayer, interview with the authors, 2020.

28. Eben Bayer, interview with the authors, 2020.

29. Arnaud de la Tour, Massimo Portincaso, Nicolas Goeldel, Arnaud Legris, Sarah Pedroza and Antoine Gourévitch, *Nature Co-Design: A Revolution in the Making*, Boston Consulting Group and Hello Tomorrow, 2021, <hello-tomorrow.org/bcg-nature-co-design-a-revolution-in-the-making/>.

30. Royal Botanic Gardens Kew, *State of the World's Fungi 2018*.

31. Francisco Kuhar, Giuliana Furci, Elisandro Ricardo Drechsler-Santos and Donald H Pfister, 'Delimitation of Funga as a valid term for the diversity of fungal communities: The Fauna, Flora & Funga proposal (FF&F)', *IMA Fungus*, vol. 9, 2018, <doi.org/10.1007/BF03449441>.

32. Patryk Nowakowski, Sylwia K Naliwajko, Renata Markiewicz-Żukowska, Maria H Borawska and Katarzyna Socha, 'The two faces of *Coprinus comatus*—Functional properties and potential hazards', *Phytotherapy Research*, vol. 34, issue 11, 2020, <doi.org/10.1002/ptr.6741>.

33. Jerzy Falandysz, 'Mercury bio-extraction by fungus *Coprinus comatus*: A possible bioindicator and mycoremediator of polluted soils?', *Environmental Science and Pollution Research*, vol. 23, 2016, <doi.org/10.1007/s11356-015-5971-8>.

34. Mengwei Yao, Wenman Li, Zihong Duan, Yinliang Zhang and Rong Jia, 'Genome sequence of the white-rot fungus *Irpex lacteus* F17, a type strain of lignin degrader fungus', *Standards in Genomic Sciences*, vol. 12, 2017, <doi.org/10.1186/s40793-017-0267-x>.

35. Dong XiaoMing, Song XinHua, Liu KuanBo and Dong CaiHong, 'Prospect and current research status of medicinal fungus *Irpex lacteus*', *Mycosystema*, vol. 36, 2017, <cabdirect.org/cabdirect/abstract/20173091977>.

36. He-Ping Chen, Zhen-Zhu Zhao, Zheng-Hui Li, Tao Feng and Ji-Kai Liu, 'Seco-tremulane sesquiterpenoids from the cultures of the medicinal fungus *Irpex lacteus* HFG1102', *Natural Products and Bioprospecting*, vol. 8, 2018, <doi.org/10.1007/s13659-018-0157-y>.

Further reading

GENERAL

David Moore, Geoffrey D Robson and Anthony PJ Trinci, *21st Century Guidebook to Fungi*, Cambridge University Press, 2011

Merlin Sheldrake, *Entangled Life: How Fungi Make Our Worlds, Change Our Minds and Shape Our Futures*, Vintage Arrow, 2021

Peter McCoy, *Radical Mycology: A Treatise on Seeing and Working With Fungi*, Chthaeus Press, 2016

Royal Botanic Gardens Kew, *State of the World's Fungi*, 2018, available online <stateoftheworldsfungi.org>

Royal Botanic Gardens Kew, *State of the World's Plants and Fungi 2020*, 2020, available online <kew.org/science/state-of-the-worlds-plants-and-fungi>

FOOD

Alison Pouliot and Tom May, *Wild Mushrooming: A Guide for Foragers*, CSIRO Publishing, 2021

David Agora, *Mushrooms Demystified*, Clarkson Potter/Ten Speed, 1986

Michael Pollan, *The Omnivore's Dilemma: A Natural History of Four Meals*, Penguin Press, 2006

MEDICINE

Christopher Hobbs, *Christopher Hobbs's Medicinal Mushrooms: The Essential Guide: Boost Immunity, Improve Memory, Fight Cancer, Stop Infection, and Expand Your Consciousness*, Storey Publishing, 2021

Mushroom References [website], <mushroomreferences.com>

Paul Stamets, *Growing Gourmet and Medicinal Mushrooms*, Clarkson Potter/Ten Speed, 2000

PSYCHEDELICS

Aldous Huxley, *The Doors of Perception: And Heaven and Hell*, Chatto & Windus, 1954

Alexander Shulgin, *The Nature of Drugs: History, Pharmacology, and Social Impact*, Transform Press, 2021

Alexander Shulgin and Ann Shulgin, *TiHKAL: The Continuation*, Transform Press, 2002

Françoise Bourzat and Kristina Hunter, *Consciousness Medicine: Indigenous Wisdom, Entheogens, and Expanded States of Consciousness for Healing and Growth*, North Atlantic, 2019

James Fadiman, *The Psychedelic Explorer's Guide: Safe, Therapeutic, and Sacred Journeys*, Park Street Press, 2011

Julie Holland, *Good Chemistry: The Science of Connection, from Soul to Psychedelics*, Harper Wave, 2020

Michael Pollan, *How to Change Your Mind: The New Science of Psychedelics*, Penguin, 2019

Terence McKenna, *The Archaic Revival: Speculations on Psychedelic Mushrooms, the Amazon, Virtual Reality, UFOs, Evolution, Shamanism, the Rebirth of the Goddess and the End of History*, HarperCollins, 1998

Terence McKenna, *Food of the Gods: The Search for the Original Tree of Knowledge: A Radical History of Plants, Drugs, and Human Evolution*, Random House, 1980

ENVIRONMENT

Annie Leonard, *The Story of Stuff* [video], <youtube.com/watch?v=9GorqroigqM>

Paul Stamets, *Mycelium Running: How Mushrooms Can Help Save the World*, Clarkson Potter/Ten Speed, 2005

Suzanne Simard, *Finding the Mother Tree: Uncovering the Wisdom and Intelligence of the Forest*, Allen Lane, 2021

Acknowledgements

This book was cast from a melting pot of eclectic and intrepid souls. Our deepest gratitude and love to all the people, stories and perspectives that left their lasting imprint on our consciousness.

To Paulina de Laveaux, we are in awe of your editorial instincts. Your belief and friendship guided us from conception through to completion. It's been an immense joy.

To Evi O, this project materialised from your faith in us. Your wickedly sharp eye (inner included) and creative vision are unmatched. Never change.

To Joana Huguenin, you brought this book to life with your biomorphic style and digital artistry through countless iterations. We are inspired by your dedication to your craft.

To Sam Palfreyman, your warmth, intellect and rigor imbued this book with clarity and cohesion. Always sweat the details – and we will too.

To Lorna Hendry, your love for words and care for every line made all the difference. You are truly a creative professional.

To Gunther Weil, your spirit and wisdom will always be a part of us and this book. We are privileged to be in your orbit.

To Wilson Leung, your design flair is woven into every inch of every page. Keep expressing you.

To Thames & Hudson, for taking a chance on our exhibit in your museum without walls.

And of course, to all the experts and mycophiles who contributed, your mission and generosity is deeply appreciated:

Alison Pouliot
Amanda Feilding
Andrew Millison
David Breslauer
David Zilber
Eben Bayer & Ecovative team
Ehab Sayed
Eugenia Bone
Francesca Gavin
Françoise Bourzat & CM team
Geoff Dann
Giuliana Furci
Gunther Weil
Jeff Chilton
Jeff Ravage

Jim Fuller & Fable team
Levon Durr
Marshall Tyler
Mary Cosimano
Maurizio Montalti
Paul Gilligan
Peter McCoy
Rick Doblin & MAPS team
Robert Rogers
Sehroon Khan
Shauna Toohey
Sophia Wang
Thomas Roberts
Tom May
Tradd Cotter

From Michael to Dad, Mum, Andrew, Juanita and Yun. Your love gives me the heart to zag.

From Yun to Dad, Mum, Lu Wen Li, Ken, Bonnie and Michael. You teach me unconditional love.

Finally, to the teacher that is this simulation. To all the synchronicities, possibilities and inspiration that have come our way as we danced through forests and cities. To all the lessons learnt so far, and to all the lessons that remain. Thank you.

INDEX

Note: page numbers in **bold** refer
to illustrations or captions.

ABOUT THE
AUTHORS

Beginning his career building technology and consumer startups, Sydney-born **MICHAEL LIM** co-founded an online eyewear brand at the age of twenty-one, which is now one of Australia's largest eyewear chains. However, early transformational experiences with psychedelics inspired his fascination with the fungi kingdom and prompted a career change. He now dedicates his time to researching fungi, psychedelics, ecology and anthropology. His exploration of the psyche and how nature gives rise to altered states of consciousness has led him on a path of self-enquiry and towards the integration of his shift in worldviews. Through his writing, Michael seeks to empower a deeper understanding of the human experience through art and science.

A gifted researcher, **YUN SHU** is dedicated to the study of consciousness and uses language and culture as tools for connection and healing. Born in Shanghai, she was exposed to the benefits of traditional Chinese medicine and fungi from a young age. After a successful career in corporate strategy in the banking sector in both Sydney and London, she turned to spiritual enquiries through Vinyasa yoga teacher training. Concurrently, her early psychedelic experiences shaped her holistic enquiry into the human experience. Now, Yun seeks to share the wisdom and knowledge she has acquired through years of practical research.

Michael and Yun's exploration has led to their philosophical enquiries, which aim to reconcile science and spirituality. They collaborate on creative projects, which are housed at Psy Earth.

www.psy.earth
@psyearth_

First published in Australia in 2022
by Thames & Hudson Australia Pty Ltd
11 Central Boulevard, Portside Business Park
Port Melbourne, Victoria 3207
ABN: 72 004 751 964

First published in the United Kingdom in 2022
by Thames & Hudson Ltd
181a High Holborn
London WC1V 7QX

First published in the United States of America
in 2022 by Thames & Hudson Inc.
500 Fifth Avenue
New York, New York 10110

The Future is Fungi
© Thames & Hudson Australia 2022

Text © Michael Lim and Yun Shu 2022
Illustrations © Joana Huguenin 2022

Thames & Hudson Australia wishes to acknowledge
that Aboriginal and Torres Strait Islander people
are the first storytellers of this nation and the
traditional custodians of the land on which we live
and work. We acknowledge their continuing culture
and pay respect to Elders past, present and future.

ISBN 978-1-760-76160-8
ISBN 978-1-760-76278-0 (U.S. edition)

British Library Cataloguing-in-Publication Data
A catalogue record for this book is available from
the British Library

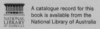 A catalogue record for this
book is available from the
National Library of Australia

Library of Congress Control Number
2021949139

Every effort has been made to trace accurate
ownership of copyrighted text and visual materials
used in this book. Errors or omissions will be
corrected in subsequent editions, provided
notification is sent to the publisher.

Design: Evi O. Studio | Evi O. & Wilson Leung
Editing: Lorna Hendry
Illustration: Joana Huguenin

Printed and bound in China by
C&C Offset Printing Co., Ltd

Be the first to know about our new releases,
exclusive content and author events by visiting

thamesandhudson.com.au
thamesandhudson.com
thamesandhudsonusa.com